MW01201114

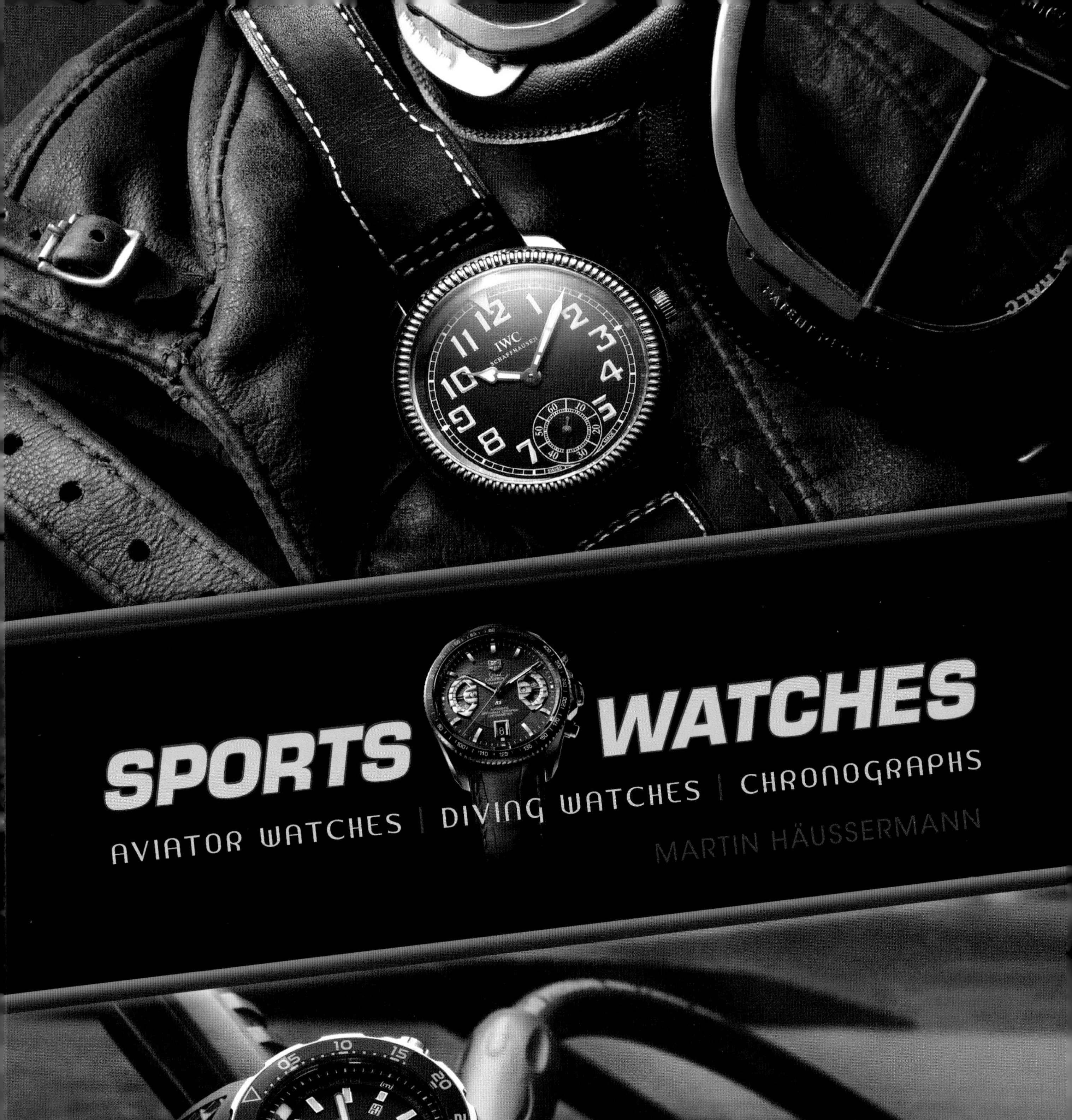
SPORTS WATCHES
AVIATOR WATCHES | DIVING WATCHES | CHRONOGRAPHS
MARTIN HÄUSSERMANN

Schiffer Publishing Ltd
4880 Lower Valley Road • Atglen, PA 19310

Other Schiffer Books on Related Subjects:
Breitling: The History of a Great Brand of Watches 1884 to the Present, $49.95
ISBN: 9780764326707
Omega Highlights, $29.99
ISBN: 9780764342127
Patek Philippe® Highlights, $29.99
ISBN: 9780764343223

Library of Congress Control Number: 2014937847

Type set in Univers LT Std

ISBN: 978-0-7643-4599-9
Printed in China

Published by Schiffer Publishing, Ltd.
4880 Lower Valley Road
Atglen, PA 19310
Phone: (610) 593-1777; Fax: (610) 593-2002
E-mail: Info@schifferbooks.com

For our complete selection of fine books on this and related subjects, please visit our website at www.schifferbooks.com. You may also write for a free catalog.

This book may be purchased from the publisher. Please try your bookstore first.

We are always looking for people to write books on new and related subjects. If you have an idea for a book, please contact us at proposals@schifferbooks.com.

Schiffer Publishing's titles are available at special discounts for bulk purchases for sales promotions or premiums. Special editions, including personalized covers, corporate imprints, and excerpts can be created in large quantities for special needs. For more information, contact the publisher.

Originally published by HEEL Verlag GmbH as *Die beliebtesten Sportuhren: Fliegeruhren, Taucheruhren, Chronographen.*

Translated by David Johnston.

PIGUET

BREITLING
1884

TUDOR
GENEVE
CHRONOGRAPH
SWISS MADE

IWC
AUTOMATIC

PANERAI
Automatic

OMEGA
CO-AXIAL
CHRONOMETER
1200 m / 4000 Ft

Contents

BLANCPAIN
500 Fathoms

What is a sports watch? A watch that is dedicated to one type of sport? Or a watch that looks sporty? Or a watch that one puts on to play sports? There is a little truth in each. Let us agree, therefore, on the following definition, which after discussions with several jewelers seems most appropriate to the author: a sports watch is a sporty-looking timepiece, usually made of steel, which is so sturdy that it remains on the wrist even while playing sports. Or, in short: the perfect everyday watch.

This an attribute that applies to almost all of the watches we illustrate in this book; however, they are very different in character. For example, there is the simple Royal Oak, with which Audemars Piguet shows that a steel watch can project luxury. Fans of sports watches will find a large number of chronometers in this book—from those with a reserved elegance to the more extravagent. For underwater sportsmen or for people who like to wear their watch while swimming, there are many selections of diver's watches. For those who choose to pay homage to the heroes of the air, there are also aviator's watches.

All of the watches illustrated here come from the twenty-first century. Some have already been illustrated in the magazine *Armbanduhr* (*Wristwatch*), although many are no longer in the manufacturers' catalogs. Nevertheless, they are still of interest, which is why we have chosen to include them. The listed prices are those that were current when the book went to press. But as the price spiral of the watch industry turns faster than any printing press, we cannot guarantee that the prices quoted will still be valid when you read this book.

Even though we have held most of the watches in our hands—and tested many of them personally—it is not our wish to make tough comparisons and purchasing recommendations. We want to whet the appetite for micromechanical masterpieces, which perhaps already are or soon will become valued everyday companions.

Martin Häußermann

PATEK
48
36
24
12
60
45
15
30

Sports Watches with Three or More Hands

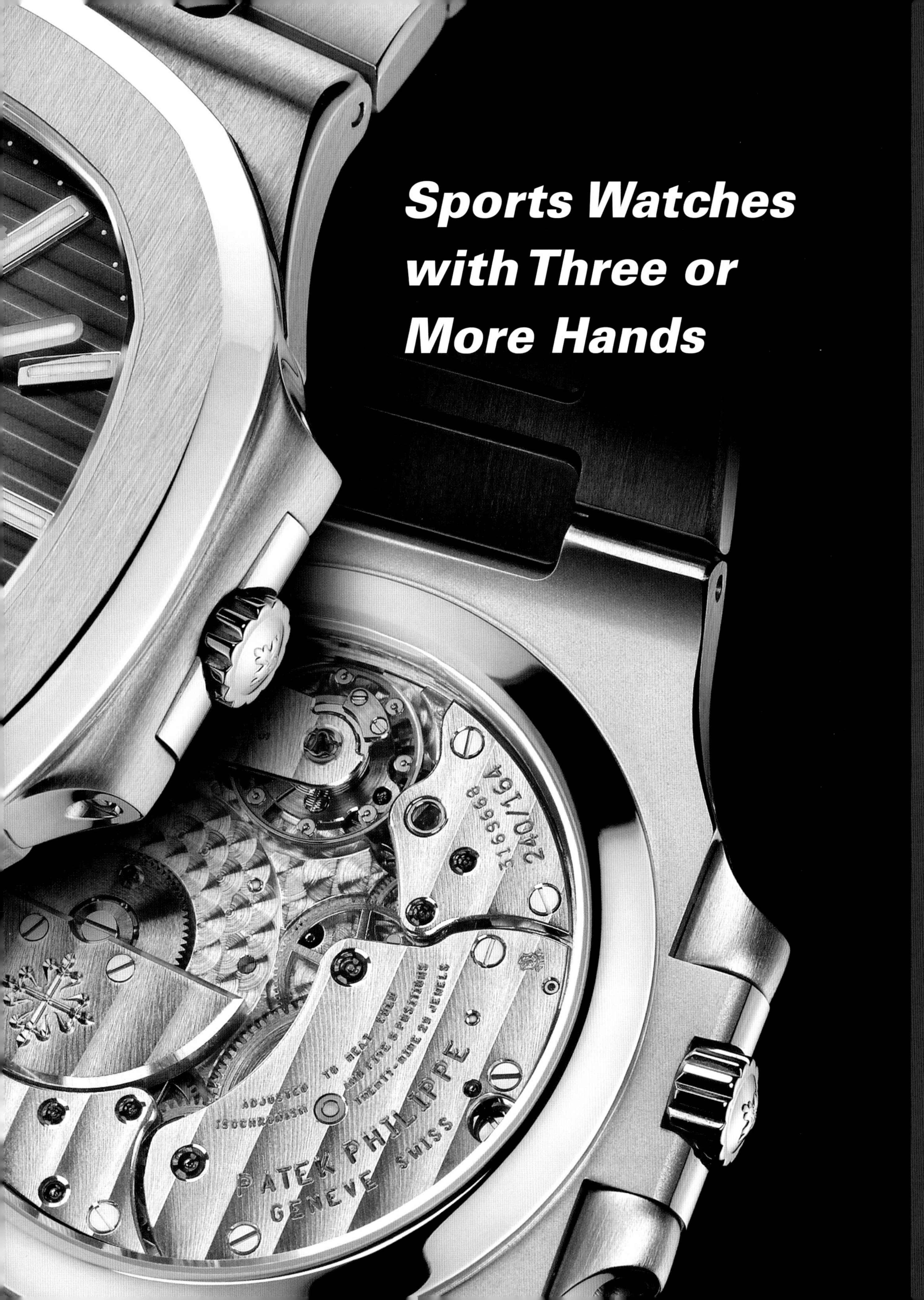

AUDEMARS PIGUET
AUTOMATIC
PATEK PHILIPPE
GENEVE
IWC
SCHAFFHAUSEN
INGENIEUR

Pure Formality

In the 1970s, the manufacturers Audemars Piguet, IWC, and Patek Philippe Furore made sports watches that all had one thing in common: they were designed by Gérald Genta. At the time he could never have imagined that his designs would remain up-to-date for more than three decades—and even in the twenty-first century would be considered the archetypes of the elegant sports watch.

Nowadays, use of the expression "classic" has become almost inflationary—it's often used for styles and products that rarely remain on the market for more than two years. As an aside, according to reference books, the term *classicus* initially meant nothing more than first-class, exemplary, and outstanding. Products possessing such qualities have long lives. Whether a product is a classic, therefore, can only be determined after a long period of time. One can thus safely describe the Royal Oak by Audemars Piguet and the Nautilus by Patek Philippe as classics; both have been on the market without interruption since 1972 and 1976, respectively. Now, as then, both are in great demand by the public. This is why those in charge of the two companies launch new variants of these successful watches at regular intervals—and in the process warm up the competition for this artistic form.

The International Watch Co. also cares for its old and successful Engineer with much love. But for the Engineer, the matter is a little different than with the Nautilus or Royal Oak. The original design dates from the year 1955, but it was just a face in the crowd until 1976, when a certain Gérald Genta redesigned it under the name Engineer SL. Of course, the Royal Oak and the Nautilus were also products of the pencil of this avant-garde designer. The idea was the same in all three cases: to design a steel watch that would be robust and suitable for everyday use, but with a touch of elegance and

The octagonal bezel secured by eight screws is a typical feature of the Royal Oak.

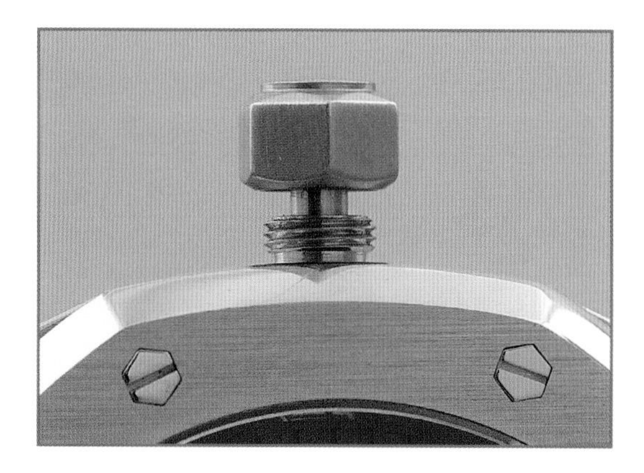

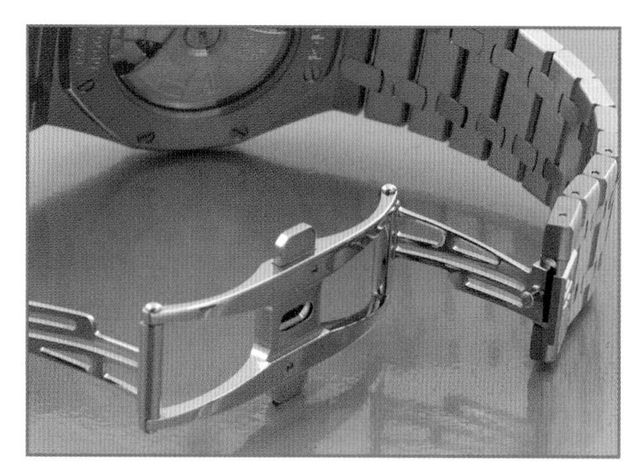

The Royal Oak is a long-runner for Audemars Piguet, which is why the basic design of the watch has never been changed.

extravagance so as to break the conservative image of the makers—to square the circle, as Genta put it.

He succeeded in this, not only stylistically but also in purely formal terms. On the Royal Oak this can be clearly seen from its prominent bezel, the contour of which consists of eight segments of a circle. They form an octagon whose interior turns into a circle. That's how it was designed in 1972 and how it has remained until the present day. The Royal Oak remains an icon among sports watches, which is why changes to the Royal Oak have always been moderate in nature. This is also true of the 2005 model, which we illustrate on these pages. Its most important characteristics: the designers decided on a three-part housing instead of the previously standard monocoque case into which the movement was inserted from above. Also, for the first time, the Royal Oak Automatic has a screw-down crown. This was previously reserved for chronographs and is partly responsible for the Royal Oak now being waterproof to 50 meters. This sports watch, which is also affectionately called "Jumbo" by AP fans, was never petite, and while it has grown little in size, it is now dimensionally up to date. The cranked band lugs ensure that the watch can comfortably be worn even on somewhat tiny wrists.

Audemars Piguet Royal Oak

Movement: Self-winding, AP-Caliber 3120; diameter 26.6 mm; thickness 4.25 mm; 40 jewels; 21,600 A/h; 60-hour power reserve.

Functions: Displays hours, minutes, central second hand, date.

Case: Stainless steel, diameter 39 mm, thickness 8.5 mm; sapphire glass; screw-locked crown, waterproof to 50 m, case back with sapphire glass insert; bezel, case center section and base joined by eight through-bolts.

Bracelet: Stainless steel with folding clasp.

Magnetic fields to 80,000 A/m. Steel case back screwed into case center-section.

Bracelet: Stainless steel with folding clasp; patented length-adjusting system.

--rists should not be too small, however, or the effect of the wide folding clasp will become counterproductive. It provides added comfort for the normal wearer, however it is rather uncomfortable on very small wrists. A test fitting by the jeweler is therefore strongly advised. The fitting also enables one to confirm that the clasp is working properly. It should close smoothly with no discernable play and open just as easily with the help of the two side fasteners. Those who like details will be pleased by the two AP logos that form the two wings of the butterfly clasp. The new bracelet is much improved over those of earlier Royal Oaks. The razor-like edges are blunted, which not only enhances wearing comfort, but also has a positive effect on the life of shirt cuffs.

Traditionally, the watch face design consisted of small squares, called *Grande Tapisserie décor* internally. The faceted illuminated hands and markers of 18-karat gold were somewhat larger and more prominent, but without deviating significantly from the original. AP only deviated from the puristic design of the original Jumbo in one respect; the new Royal Oak has just one second hand.

IWC Engineer Automatic

Movement: Self-winding, IWC Caliber 80110; diameter 30 mm; thickness 7.25 mm; 28 jewels; 28,800 A/h; 44-hour power reserve; integrated shock absorber.

Functions: Displays hours, minutes, central second hand, date.

Case: Stainless steel, diameter 42.5 mm, thickness 14.5 mm; sapphire glass; screw-in crown, waterproof to 120 m; inner cage of soft iron shields the movement against magnetic fields to 80,000 A/m. Steel case back screwed into case center-section.

Bracelet: Stainless steel with folding clasp; patented length adjusting system.

"Steel outside and gold inside." This slogan from the birth of the Royal Oak is still true today. For its 33rd birthday, it received a special gift: the 3120 automatic movement, first shown in 2004, which, as the advertising slogan suggests, had a gold rotor (engraved with the coat of arms of the Audemars and Piguet family) that could be admired through an aperture in the case back. This perfectly finished movement replaced the Caliber 2121, which, though popular with fans, was somewhat dated. It was based on a design by Jaeger-LeCoultre. The new watch was made somewhat thicker than its predecessor, as the designers believed this would make it more sturdy. Nevertheless it is not lacking in elegance and the new Royal Oak is still amazingly thin for a sports watch.

One cannot really say the same of the IWC Engineer. Its housing measures all of 14.5 millimeters in depth and 42.5 millimeters in diameter, surpassing all of its predecessors. The Engineer is a piece of mechanical engineering that became a watch. Such a striking piece of stainless steel does not want to be elegant. Instead it flaunts its message: I will never let my wearer down. And one believes this of it, not just because of its appearance. It also has various design features that ensure reliable operation even under adverse conditions. In keywords: waterproof to 50 meters, magnetic field protection to 80,000 A/m, and an improved shock-resistance system.

To understand why IWC made the Engineer an horological treasure in the year 2005, one must take a look back into the history of this model series. The first Engineer, which came on the market in 1955, was simply a development of the Mark XI aviation watch that was built for pilots of the Royal Air Force in 1948. It already had the inner case of soft iron, which screened the mechanism from magnetic fields and thus eliminated their negative effect on chronometric precision. But the people in Schaffhausen reasoned that pilots weren't the only ones who needed an accurate watch and worked in spaces influenced by magnetic fields. The same applied to engineers. So they created a simple round men's watch for this professional group and gave it the brand name Engineer. The shape of the original Engineer was anything but spectacular, which is why, in the mid-1970s ,IWC tasked Gérald Genta with finding a new form. The result was the Engineer SL, unveiled in 1976, which has since become a design icon. While classics are often ahead of their time, few are recognized as such when they are first introduced. Only about 1,000 examples, around 600 of them with a mechanical movement, were made. In the early '80s, fashion and the quartz wave created a demand for ever thinner watches. And so

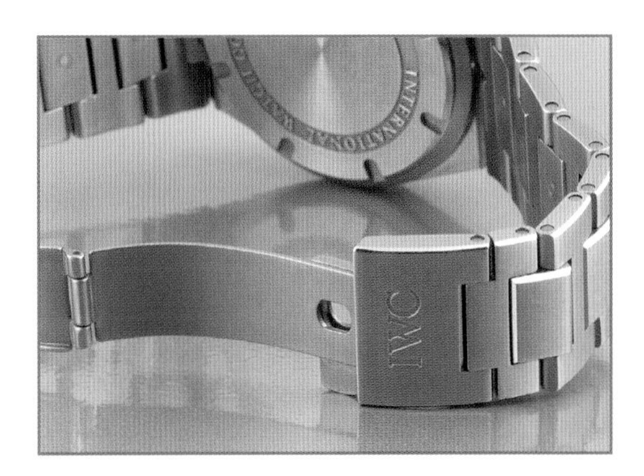

With its striking appearance, the IWC Engineer embodies robustness and dependability.

Patek Philippe Nautilus Gangreserve

Movement: Self-winding, PP Caliber 330 SC IZR; diameter 27 mm; thickness 3.5 mm; 30 jewels; 21,600 A/h; 48-hour power reserve.

Functions: Displays hours, minutes, central second hand, date, power reserve indicator (patented winding indicator).

Case: Stainless steel, two part, diameter, from the 3 to the 9, 42 mm without crown, from the 6 to the 12, 44.5 mm, thickness 8.1 mm; sapphire glass; screw-down crown, waterproof to 120 m.

Bracelet: Stainless steel with folding clasp.

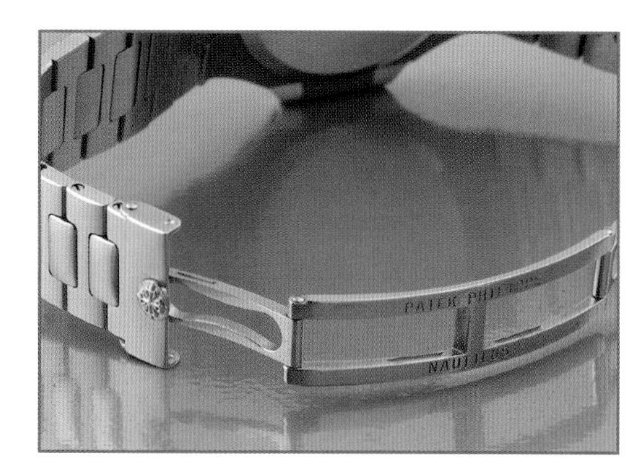

In the Nautilus, Patek Philippe made the complete everyday luxury watch. It has since grown into an entire family of models.

the Engineer also became ever thinner, but without abandoning Genta's concept. Following in the footsteps of the Royal Oak and the Nautilus, the Engineer's steel bracelet was integrated into the housing, as a result of which both components became an organized whole. Also absolutely typical for the Engineer was the bezel screwed into the housing with its five holes. While these were initially design features, they could serve as starting points for a case opener.

Because the '70s are in vogue again, and because the Gérald Genta design can be regarded as style-shaping for this model range, in configuring the Engineer for the new millennium, IWC designers Guy Bove and Matthias Kummer used it as their inspiration without copying it. Of course the typical bezel had to stay, but it was no longer held in place by screws. That flew in the face of IWC's new sense of aesthetics. It often happened that the upper hole was just in front of or just after the 12. "That looked somehow lopsided, therefore we considered another solution," explained IWC man Pius Brida. Simply press-fitting the bezel was out of the question, therefore the glass rim was attached to the housing by a bayonet lock. The bezel was fitted and then rotated ten degrees. Then not only was the housing sealed, but the upper hole was precisely above the 12. The steel caseback, on the other hand, was screwed into the housing center section like that of its predecessors. In combination with the screw-down crown, this made the watch waterproof to 120 meters.

The horological purists were pleased, but the aesthetes complained that the mechanism was hidden behind a massive steel base. A transparent back cover made little sense for the Engineer, for it only gave a view of the soft iron inner housing. This consists of a ring, a pressure cover, and the watch face. This effectively shields the mechanism against magnetic fields with strengths up to 80,000 Ampere/meters (A/m). The Swiss standard requires "anti-magnetic" watches to have a protection of 4,800 A/m. Put simply, magnetic fields can magnetize metal watch components, especially the escapement parts, and cause inaccuracies up to stopping the watch. Of course that would not do on a watch called the Engineer. This puristic way of thinking has gone out of vogue at IWC. For example, for a long time there have been Engineer watches with glass bases—and without magnetic shielding.

The designers worked hard not only on the case of the Engineer Automatic, but also on the movement. It is called Caliber 80110 and may be characterized as a manufacturer automatic movement even if components do come from the ETA cupboard. The Caliber 80110 borrows from the Caliber 5000 family (from the Portuguese family) but without simply being a miniaturized version. "More accurately, it is a further development," observed designer Denis Zimmermann. "We completely redesigned the Pellaton winding system." This efficient winding system, which bears the name of the former IWC head Albert Pellaton, is now substantially more robust than earlier designs. As well, there is a shock-absorbing system consisting of an S-shaped automatic bridge and a partially milled gyrating mass. Thus, according to the press release, the Engineer is "well equipped to meet all earthly requirements, in everyday use as well as sport, even in its most extreme form."

That is a good thing. For despite its weight of about 200 grams, the Engineer is so comfortable that one simply doesn't want to take it off to play sports. This is due in large part to the outstanding steel link bracelet, which fits easily onto the arm circumference, and the simple yet very functional folding clasp. As a result, it is not only in sports that the Engineer cuts a good figure. It can definitely be worn with a suit, although one should choose cuffs that provide sufficient room.

The wearer of a Nautilus by Patek Philippe, however, really need not worry about such things. At about eight millimeters, the sports watch from the Geneva luxury manufacturer is significantly thinner than the IWC and thus slips easily under any cuff. The cuff is not placed under any strain, thanks to the gentle curves of the housing and the perfectly trimmed corners of the bracelet. The supple link bracelet and the butterfly clasp ensure perfect wearing comfort. But, if you will pardon my saying so, a buyer expects nothing less from Patek.

The Steel Nautilus is something of an orchid in the Patek Philippe collection. It is sold only in limited quantities. Co-owner and managing director Thierry Stern explains this thusly: "Because we have only a limited capacity in watches, we of course prefer to enclose them in housings of gold or platinum." In 2005, Patek Philippe granted friends of the brand a new variant: the 3712/1 with a power reserve indicator, seconds hand and moon-phase indicator. For reasons of better comparability we have used the somewhat simpler 3710/1A, which is no longer in production but which is still available on the collector's market.

Audemars Piguet proudly displays the self-winding manufacturer Caliber 3120 through an aperture in the case back.

And so it presented itself in 1998—after more than ten years—again in the big housing of the original Nautilus, which, like the Royal Oak and the Engineer, answered to the name of Jumbo. Incidentally, the Nautilus, like the Engineer

In its day, the Nautilus test watch showed its dignified elegance through Roman numerals on the dial. Today the watch is more reserved, with plain markers.

SL, celebrated its premiere in 1976, but in character, it is more like the Royal Oak, launched four years earlier. Here again the characteristic styling element is the bezel. The edge of the glass consists of eight circular segments, which together form an elongated octagon. The term bezel is thus imprecise, for the design of the steel case is unique. The Patek movement is installed in the one-piece case from above and is closed from above in the manner of a porthole. On the left of the housing is a massive hinge, on the right the "porthole" is closed with equally massive screws—compressing the rubber seal between the upper and lower parts. The Nautilus is thus watertight to 120 meters and honors—at least to some extent—its namesake, the nuclear submarine, launched in 1955.

A dial could scarcely be tidier than that of the IWC Engineer.

In contrast to that of the Royal Oak, the Nautilus's glass, which has minimum overlap, also takes the octagonal shape of the bezel, which underlines the unique face of this watch, but which is also very complicated and thus expensive to manufacture. Unfortunately the designers neglected to add a viewing aperture in the case back. Thus, for the most part, the Caliber 330 SC IXR works furtively. A pity, for the automatic movement is marked with the official seal of Geneva and is not just an optical delicacy. It also has a patented indicator that informs the wearer about the forces in the mainspring. Patek Philippe calls the system, which consists of a rotating disc and an overlying pointer, a "winding zone indicator." It functions as follows: the disc with the imprinted comet's tail is linked to the winding system and turns when force is transmitted to the mainspring. The pointer, on the other hand, moves with the running check spring. Both components move clockwise. When the check spring runs, the pointer moves towards the tail of the comet. If the pointer is pointing at the comet's tail, then it is high time to wind the watch, either manually or by moving the winding rotor. The indication would be more positive with a somewhat longer pointer that traverses the comet's tail.

Our greatest criticism: the Nautilus lacks the stop-seconds feature usually present on highly-accurate sports watches. And, in the opinion of the authors, all three brands of watches tested would benefit from a somewhat larger date indicator, which could have been added to the two new designs by IWC and AP, in particular, without excessive added cost. But that is the extent of our criticisms. The finish of all three timepieces befits their standing, both in terms of the movement and the case and bracelet. Readability is also outstanding thanks to one-sided anti-reflective sapphire glass and high-contrast dials. What is left is the price. Even used collector's items are not exactly bargains among sporty three-hand watches.

On the other hand, that is also an indication of how well the watches retain their value: in ancient Rome the members of the wealthiest class (*classis prima*) were called *civis classicus*. Anyone from this class would have no problem affording the modern interpretations of a Gérald Genta design by Audemars Piguet, Patek Philippe, and IWC. But even those with an interest in watches who spare themselves such a timepiece know: the quality of these watches—as previously mentioned—is unquestionable, and when one buys one, one is buying a true classic.

ROLEX
SKY-DWELLER
SWISS
MADE
28

King of the Sports Watches

The Rolex name stands for luxury, quality, reliability—and for sportiness. Rolex watches have been to the top of Mount Everest and the depths of the Marianas Trench. But the brand is also widely loved on the tennis court and racetracks. Behind it is an organization which, thanks to great production depth, controls quality from the very beginning—and that has paid off handsomely.

Even revolutionaries know a good watch. The picture of Ernesto "Che" Guevara with the Rolex Oyster on his wrist went round the world, and it made the Cuban guerilla fighter and politician an unofficial sponsor of the brand with the crown, posthumously. Once, there was much joy in Geneva, when, for example, Winston Churchill, Konrad Adenauer, and Charles de Gaulle appeared in public wearing Rolexes. Nowadays, testimonials are more likely to come from the fields of science and sport. Former champions, like racing driver Sir Jackie Stewart, golfer Arnold Palmer, and skier Jean-Claude Killy, advertise for the brand with the crown, as do active athletes, for example, tennis ace Roger Federer and skier Lindsay Vonn.

One expects no less from a brand that, like no other, stands for the luxurious sports watch—though definitely not for revolution. Continuous evolution and calculated business policies distinguish the company that is privately held and overseen by the Hans Wilsdorf Foundation, named after the founder of Rolex. Its goal is the maximum possible production depth while controlling costs and quality from the outset. Rolex also strives to be independent of suppliers.

Managing director Patrick Heiniger took a major step in this direction in 2004. He acquired the then-independent manufacturer Montres Rolex SA, Biel, and integrated it into Rolex SA, Geneva. In doing so he killed two birds with one stone: he took control of an important supplier and at the same time secured what was probably the company's most important export market. Rolex Biel made the Caliber 3135 automatic movement and its derivatives, which were used in the various Oyster models that had made the company big. As well, Rolex Biel held the marketing rights for the USA.

Until then, the Biel company had belonged to the Borer family, descendants of Jean Aegler, whom Rolex founder Hans Wilsdorf had formed ties with as his long-term, exclusive supplier in 1920. According to reports in Swiss business magazines, the supply contract was to

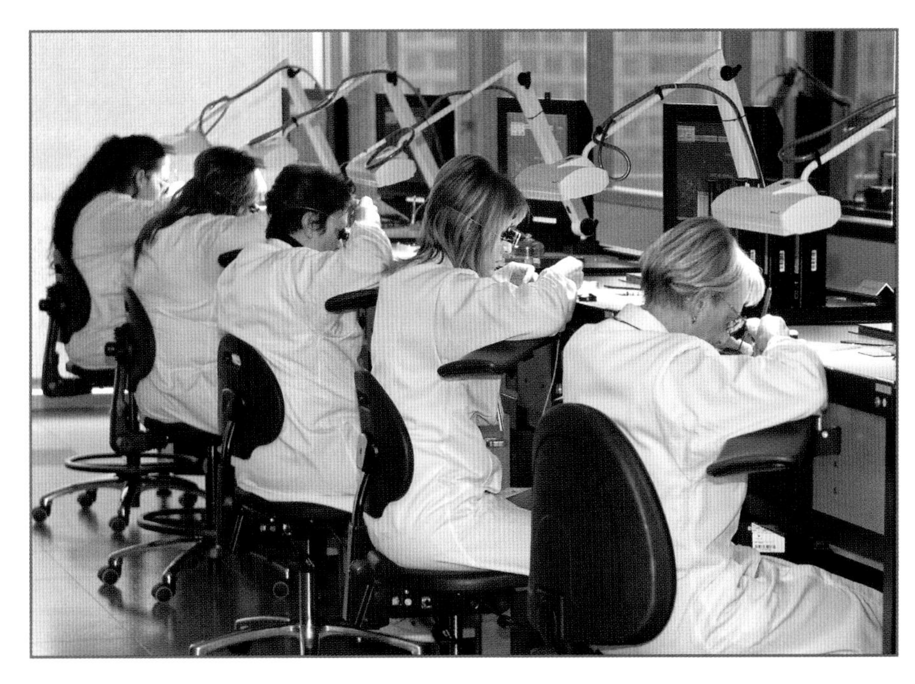

With modern production facilities and a large qualified work force, Rolex is able to produce high-quality watches in large numbers.

For a long time the Rolex company was shut tight as an oyster, which is what the Swiss named their waterproof watches. The Geneva watchmaker now allows itself to show some of its cards.

expire in 2013, which could have severely shaken the major player, whose annual sales were estimated at 3.5 billion Swiss Francs. By integrating Rolex Biel—the sale price of which was estimated at 2.5 billion Swiss Francs (about $2.8 billion at current exchange rates)—the flagship of approximately 300 Swiss watchmakers sailed into calmer waters. Precise figures have not been revealed. Here, the company keeps a low profile and does not publish financial details, as it is a private foundation. A Rolex spokesperson commented on the acquisition: "It amounts to an epochal step, which will strengthen the company in the long term."

Further steps were the construction of new—and the modernization of existing—production facilities in Geneva. In 2000, the facility in Chêne-Bourg was opened after a construction period of two years. Rolex has concentrated all its activities in the field of development and manufacture of watch dials and gem-setting there, with a production area of approximately 13,000 square meters. A fully-automatic high-bay warehouse provides the production floor with suitable materials as they are needed. Systems ensure that production does not stop even if there is a problem in a storage silo. Logistics in the two other Geneva facilities follow the same concept. The company's buildings with their huge glass façades provide an example for other producers to follow, and not just architecturally.

Take Plan-les-Ouates, for example. There, in the same neighborhood as Patek Philippe, Vacheron Constantin, and Piaget, Rolex produces cases and metal bracelets on eleven floors—five of them below ground. The operation encompasses not only assembly, but also development, design, and quality control. Rolex employees even smelt some of the gold needed for cases and bracelets in an in-house forge, dressed in helmets and silver fire-protection coats like the legendary firefighter "Red" Adair. Another example set by Rolex is "Everose," a patented alloy of 76 percent pure gold, 22 percent copper and two percent platinum. This mixture, thanks to

Rolex in Brief

1905: Hans Wilsdorf of Kulbach (Bavaria) establishes a business in London specializing in the sale and distribution of watches.

1908: Wilsdorf creates the Rolex brand.

1920: Wilsdorf establishes Montres Rolex SA, which provides movements exclusively to maker Jean Aegler of Biel.

1945: The childless Wilsdorf establishes a foundation and places his Rolex shares in it.

1960: Hans Wilsdorf dies; his longtime associate André Heiniger takes charge of running the company.

1992: André Heiniger installs his son Patrick as CEO.

2004: Rolex Geneva buys the Rolex Biel operation from the Borer family (descendants of Jean Aegler).

2005: The production facility in the Plan-les-Oates industrial district of Geneva is opened.

2006: Rolex manufactures 613,000 tested chronometers and generates an estimated sales volume of three-million Euros.

2007: Rolex produces at four facilities in Geneva: Acacias (headquarters, F + E, assembly), Plan-les-Ouates (bracelet and case manufacture), Chêne-Bourg (dial manufacture, diamond trimming), and Biel (movement production; the Caliber 3135, centerpiece of the Rolex collection, is created there). Rolex employs 6,000 people in Switzerland and 8,000 worldwide. The watches are sold globally by more than 28 subsidiaries.

2008: Former financial director Bruno Meier replaces Patrick Heiniger as CEO.

2011: Riccardo Marini, former head of Rolex Italia, becomes the new CEO.

the copper, has a red coloring and is more durable than common commercial alloys (75 percent gold, 21 percent copper, four percent silver) that literally yellow over the course of many years. The Geneva company buys its steel from the Austrian maker Böhler. It is of course of the highest quality, designated by the number 904L, and is also used to manufacture surgical instruments. Owners of the rare steel Rolex need not fear rust, even if the watch must regularly be immersed in water.

In another area of production, oxide formation is not frowned upon, but rather it is desired—in the making of hairsprings. Rolex is one of the few watch manufacturers to have gained the know-how to make these tiny parts, which contribute significantly to a timepiece's accuracy, itself. Only a few other makers, such as Patek Philippe, Ulysse Nardin, Lange & Söhne, and Nivarox, for example, have mastered this technology. A company in the Vallée-de-Joux supplies the vast majority of watchmakers and is part of the Swatch Group, which also makes watch movements (ETA) and complete watches (Breguet, Blancpain, Omega, etc.). In-house hairspring production is a strategic advantage for a watchmaker that cannot be underestimated, least of all in terms of quality. The foundation of this is 280 engineers, technicians, and watchmakers who, according to a Rolex spokesperson, apply for five to eight patents per year.

Rolex also mixes the material for its so-called "Parachrom" hairsprings itself. This consists of iron, niobium, and zirconium and initially appears as an unassuming metal rod about 30 centimeters long. Not only is the alloying of the basic materials patented, but so is the special production method for which Rolex has developed a machine that is unique in the world and is no larger than a clothes dryer. In general terms, it is a vacuum chamber in which, with the help of great tension, sections of the rod are slowly heated and then cooled. This treatment results in the alloy becoming completely paramagnetic,

The Caliber 3187 is a self-winding movement with a second time zone and is used in the Explorer II. The shimmering blue Parachrom hairspring is one of the identifying features of the new generation of movements.

meaning that magnetic fields have no effect on the watch mechanism's accuracy. The previously-mentioned protective oxide layer is also formed in the process. Numerous rolling and pressing processes turn the 30-millimeter-long rod into about three kilometers of spiral spring, precisely 50 microns (.05 mm) thick. That is significantly thinner than a human hair, but unlike a hair it is the same thickness from front to back—for controlling the precision for which Rolex watches are known.

The hairsprings are made in Acacia, a suburb of Geneva. The Rolex headquarters, which was thoroughly modernized from October 2002 to October 2006, has been located there since 1965. As in Plan-les-Ouates and Chêne-Bourg, steel and glass now dominate the industrial building, in which the central company decisions are made and the entire enterprise is administered. Customer service and final watch assembly are also located there, and mechanisms for the Yacht Master II and Prince models are assembled there as well.

All of this is done in an extremely collaborative manner. For example, the Caliber 4160 movement for the Yacht Master II regatta watch passes through 60 different stations. Up to eight different assembly processes are carried out at each work station. Although there are 100 people sitting at long work tables, concentrated silence reigns. There is not much to be heard except for the hum of the small electric screwdrivers and the soft clicking of the transport system designed especially for Rolex. It consists of dustproof plastic containers that hold the movements from the first production stage to the final checks. They are, of course, provided with barcodes, so that

The new Explorer II with 24-hour hand and separate adjustable hour hand is especially popular among globetrotters. It is also available with a black dial.

the progress of each mechanism can be tracked as assembly progresses. The room can only be entered through air locks and workers must wear special shoes—visitors are given overshoes—and work coats. Clean room conditions prevail and permanent overpressure in the production room ensures a dust-free environment.

More than 613,000 watches were produced in this way in 2006. This number was not released by the company itself; rather, it was published by the Contrôle Officiel Suisse des Chronomètres (Official Swiss Chronometer Testing Institute) or C.O.S.C. The institute tests watch movements for accuracy. For many watchmakers, this testing is the exception; at Rolex it is the rule. Approximately 6,000 people work at the four Swiss locations. By comparison: A. Lange & Söhne makes approximately 5,500 watches per year with 450 employees. Admittedly the comparison is a little misleading, for while Rolex watches—with the exception of the Daytona and Yacht Master II chronometers—essentially give the date and time, most Lange watches are equipped with complicated secondary functions. And yet Rolex's direction is clear: the highest quality mass production.

Nevertheless, contrary to what one might expect, one searches in vain for automated production lines here. Also absent are the grey-haired watchmakers that so many Swiss or Saxon luxury watch manufacturers like to portray to the public. Anyone who buys a Rolex must get used to the idea that it simply wasn't made by his personal watchmaker in a romantic farmhouse. Instead, many people work in modern, multistory buildings to make sure that a Rolex later ticks reliably on the wrist of its owner—whether in everyday use or in sports.

30
15
MADE

From Time to Time

While travelling, it is advantageous to have a watch that can quickly be set to the new local time without losing home time. Steel cases and bracelets give the horological globetrotter sporty elegance and manufacturer exclusivity. A look back to the year 2007.

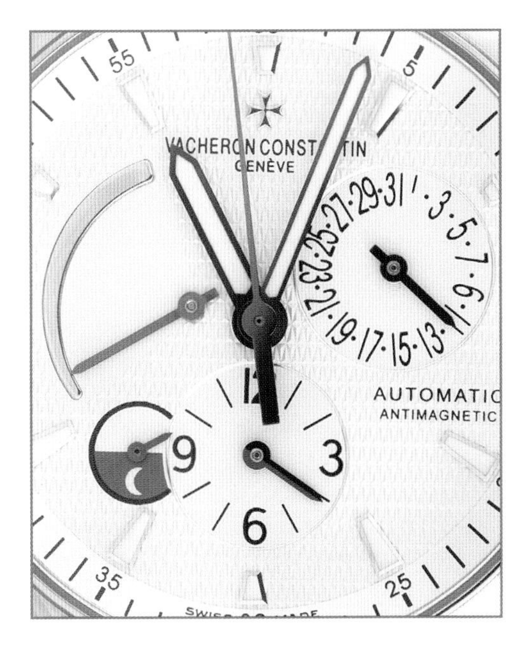

The dials of the GMT watches illustrated here are sometimes minor works of art.

The German poet Mathias Claudius once said that when one makes a journey, one has tales to tell. In most cases, those who make a journey also have to keep track of two times: local and home time. After all, our world is divided into twenty-four time zones that are numbered east from the prime meridian because of the earth's west-east rotation. As one travels, the local time changes both east and west, by an hour for every 15 degrees of longitude—at least, theoretically. The earth turns once per day (360 degrees) and the day consists of 24 hours (360 divided by 24 is 15).

Great Britain laid the foundation for this calculation by establishing the zero meridian. For while the zero parallel, or the equator, is clearly defined geographically, the zero meridian can only be defined arbitrarily. This was done by the Royal Observatory in Greenwich, a suburb of London, in a nautical almanac that appeared in 1767. The English decided that, as a baseline for measuring world time, the zero meridian should run through their capital city—precisely through the observatory. This was confirmed by the International Meridian Conference in 1884, and consequently the reference time became known as Greenwich Mean Time or GMT. Today the term "Universal Time Coordinated" (UTC) is more commonly seen than GMT and is used as reference time by the military and in the field of aviation.

Germany now falls under Central European Time (CET). It was introduced in 1893 with Paris as its reference city and is equivalent to Greenwich Time plus one hour. For political and practical reasons, it extends from the west coast of Spain to the eastern border of Poland, thus covering a stretch of almost 35 degrees of longitude. This does not refute the original calculation, it merely relativises it.

But this should not concern us, especially as we are watch enthusiasts. For this situation with time zones provides us with a sound reason for involving ourselves with a special sort of timepiece. And when we make a purchase, it can be not just for the sensual experience of having a new ticker on our wrist. We can also make the argument—for example to our spouse—that the new piece really is practical and useful. At least from time to time.

The IWC Spitfire UTC is an elegant variant of the classic UTC pilot's watch. For optical reasons IWC moved the cutout for the digital 24-hour display to the lower half of the dial.

IWC Spitfire UTC:

Say Hello to the Sun

The watchword at IWC was "ready for takeoff" in 1999, when the "UTC Pilot's Watch" was first cleared for takeoff. The indication of a second time zone is a perfect match for a pilot's watch. After all, pilots often traverse continents and time zones in rapid succession. And even in their home stomping grounds, for pilots there is also the difference between local time and UTC. The IWC watchmakers therefore decided to equip the powerful and reliable ETA 2892 automatic movement with this additional hand and place it in a 39-millimeter case, whose design fit perfectly into the pilot's watch family (Mark VII pilot's chronometer and double chronometer). Of course, the pilot's watch is also shielded against the influence of magnetic fields, which have a negative effect on accuracy. The protective element consists of the dial, a movement retaining ring, and a dust cover over the movement—all made of soft iron.

The entire design is still up to date, including the simple-to-operate added function. If the crown is unscrewed, it is carefully moved into the first detent. This allows the hour hand to be adjusted forwards and backwards. The minute hand is unaffected and the movement also continues to run. Only the disc of the 24-hour hand (home time or UTC) moves slightly. Thus not a second is lost when setting the watch to the new local time. The date indicator moves in either direction if the hour hand moves more than 24 hours, a handy feature for trips across the date line. The wearer also uses this function if he wishes to correct the date, for example in months with 30 days. A classic date change using the crown will not be found here.

The dial has changed greatly since the first takeoff, however. The UTC is no longer available as a classic pilot's watch with a black-white dial, to the regret of more than a few fans of the brand and of watches with excellent readability. When it was launched in 2003 the IWC bosses added it to the so-called "Spitfire" collection, named after a famous British fighter aircraft, hoping to attract a younger group of consumers. With superimposed markers and numbers, which, like the prominent hands, are provided with plenty of luminous material, the silver-galvanized, three-dimensional dial literally sparkles. This at least ensures excellent readability in the dark. Finally, in the most recent facelift in 2006, the 24-hour hand was moved from the 12 to the 6 o'clock position. This gives the dial a more balanced appearance, and the watch looks a little like it's smiling.

Jaeger-LeCoultre Master Hometime:

Business Class

"There are two kinds of time zone watches," lectured Janek Deleskiewicz, designer with Jaeger-LeCoultre, at the unveiling of the Master Hometime in 2004. Specifically, there are those for homebodies and those for globetrotters. The Hometime clearly fits into the latter category, because it can quickly be set to the time at the destination city and also shows the local date.

It is difficult to argue with the Frenchman. A slight tug on the crown to the first detent, and the hour hand can easily be moved forwards or backwards, with the date changing when the day boundary is crossed. A skeletonized, tempered blue hand shows the home time and is normally hidden beneath the lance-shaped hour hand. The "home hand" is coupled to a day-night indicator. This allows the traveler to avoid mistakenly waking his colleagues or loved ones at home. If the stylized moon is moving across the sector of the circle, the observer knows that the second half of the day (between 6 P.M. and 6 A.M.) has begun, while the sun indicates the period between 6 A.M. and 6 P.M.

This indicator with the sun and moon at the two ends is the only playful touch the designer allowed himself. Otherwise the Hometime is typical of the watches in the Master Line: simple and elegant, business class on the wrist. The dial appears tidy and, on closer examination, also finely worked with a highly-polished surface and superimposed markers and numbers. Like the hands, they are made of stainless steel, which does not provide sharp contrast with the silver galvanized dial. In poor lighting conditions, this reduces readability a little. The minimal luminous material on the hands and the twelve small luminous dots provide only a slight improvement. Thanks to a large aperture in the dial, the date indicator is very easy to read. There doesn't always have to be an oversized date.

The Jeager also shines with intrinsic inner values, in keeping with the Automatic Caliber 975, which was also introduced in 2004. The thoroughbred manufacturer caliber has the nickname "Autotractor," which sounds neither charming nor elegant. But a look through the viewing aperture in the base shows the mistake of such interpretations. It is large and powerful, as the name is meant to suggest, but thanks to the finest workmanship, it is noble through and through. The watch makes traveling a pleasure.

Jaeger-LeCoultre displays home time with the blue hour hand, which is coupled with the day-night indicator. The hour hand for local time can be set independently and at home covers the blue hand.

Ulysse Nardin positions the 24-hour display at the 9 o'clock position; the date display is at 2 o'clock.

Ulysse Nardin Dual Time Maxi:

Cruise King

Ulysse Nardin made a name for itself as a maker of marine chronometers soon after its founding in 1846. Numerous winners' certificates from chronometer contests in various observatories in the world attest to this, as does the anchor still present in the company's logo. When passing through multiple time zones, both seamen and passengers can make good use of a watch with a GMT function—for example during a cruise or an Atlantic crossing.

All of the features of the Dual Time Maxi are eminently suitable for such a purpose. This begins with its size. A 42-millimeter case allows for large hands, easily read by even the shortsighted. The designers from Le Locle also gave the watch a large date indicator. Only the hour display at 9 o'clock sometimes requires a second look in order to read it correctly. The reason: this indicator's disc turns several millimeters beneath the dial surface and, because of the stout construction of the window, can be partly covered at certain viewing angles. On the other hand, this porthole reveals the love of detail, as do the finely made, wedge-shaped hour markers. Their blue flanks contrast with the silver dial.

The blue strap of crocodile leather also underlines the watch's maritime character. It is heavily padded in the center in order to maintain the proportions to the case. It only requires several weeks for the stiffer part of the strap to adjust to the wrist. Befitting its status is the fine folding clasp, which is released by applying pressure with the thumb and index finger on the two pushbuttons on the sides. For those who wish to also wear the watch in the cruise ship pool, a stainless steel bracelet is available for an extra $830 or so. Its waterproof characteristics (to 10 ATM, or atmospheres) allows the watch to stay on the arm while swimming.

Adjusting the hour hand to local time could not be simpler. Two triggers on the left flank of the case serve as tools. To avoid any error, they feature plus and minus signs. If the hour hand passes the date limit, the large date indicator automatically follows. Ulysse Nardin developed this mechanism in 1994 and also employs it in the finely finished Caliber 2892 automatic movement. It runs so precisely that you will definitely never be late for the captain's table.

Rolex GMT Master II:

One for Everyone

It makes no difference whether you travel by rail, by car, by aircraft, or by ship—or whether the trip demands a dark suit or allows for T-shirt and jeans. With the Rolex GMT Master II one is always properly dressed, especially if the watch appears in simple stainless steel clothes like our test watch. Its presence is an indication of the wearer's consciousness of quality, especially in the watch's revised design.

The conservative politicians in the House of Rolex have left the model much as it was and only changed the watch where they thought it necessary. Numerous detail improvements were made. One is the scratchproof and UV-resistant ceramic bezel. Its engraved numbers and markers are flushed out with platinum, so that even the purchaser of a steel watch gets to enjoy some precious metal. As well, the bidirectional rotating bezel rests cleanly at hourly intervals. By the way, dear collector, the days of the two-color bezel (e.g. blue and red) are numbered.

Instead of the Twinlock winding crown previously used, the designers of the GMT Master gave it the Triplock crown, which has an additional gasket ring, as used in the Submariner and Sea Dweller diver's watches. The claimed waterproof rating remains 10 atm, however. Rolex made a real advance in quality with respect to the bracelet. While that of its predecessor was criticized for being "rickety," the new bracelet, dubbed the Oysterlock, is sublime beyond any doubt. There is no longer any play in the bracelet links and the steel clasp is no longer made of canted steel but instead is milled from a single piece. It now opens and closes smoothly.

Everything as usual on the Rolex: the 24-hour hand with arrowhead pointer and the date under the characteristic magnifier.

A very useful detail is the integrated strap extender, which, at about five millimeters, is sized to compensate for the wearer's different wrist circumferences in summer and winter.

The inner workings of the watch were also changed somewhat. While it is based on the Rolex Caliber 3135 standard automatic movement, it is now called 3186 instead of 3185. The essential difference concerns the motion regulator, more precisely the hairspring. Used here is the blue Parachrom hairspring, which Rolex developed itself and also manufactures in-house. According to Rolex, the alloy used is completely anti-magnetic and considerably more shock-resistant than that used in other commercially available hairsprings. Like all Rolex watches, the GMT Master II is chronometer tested.

The GMT function has always been controlled by the crown on Rolex watches and is an example for many other manufacturers. In the first detent, the hour hand can be moved forwards or backwards independently, without holding the watch. If it is moved past 12 A.M., the date also changes. Traditionally, a 24-hour hand with large arrowhead indicates home time. This is no longer bright red, but instead is in the less obtrusive house color of dark green. Well, have a pleasant trip.

Vacheron Constantin Overseas Dual Time:

Let the Others Travel

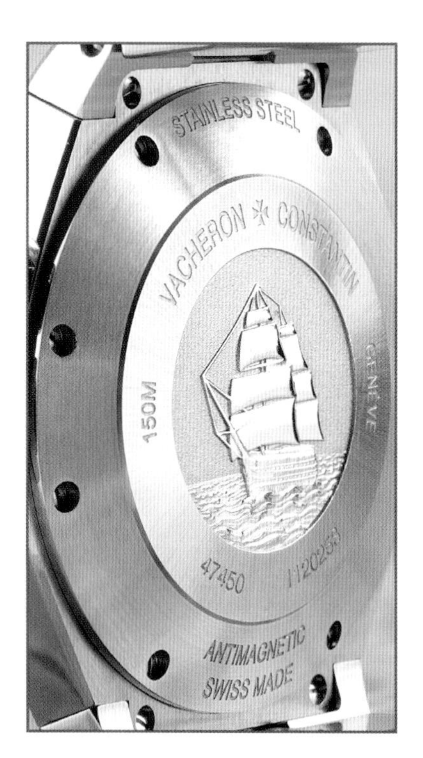

Even on the Overseas sports watch, Vacheron Constantin knows what the customer expects: class and elegance.

As a rule, those who can afford a watch from the luxury brand Vacheron Constantin get one despite the availability of special offers elsewhere. The watchmakers from Geneva ask about $16,000 for the Overseas Dual Time, making it more than twice as expensive as the Rolex. What does the buyer get in return?

Objectively one first sees a steel watch with steel bracelet that is finely made down to the last detail. But we expected nothing less. Examples required? There would be, for example, the steel bracelet, whose links bear the shape of the Maltese Cross—the brand logo. These links are joined by screwed-on bracelet holders in such a way that there is only minimal play. On the wrist, this bracelet is a true skin flatterer that catches no hairs. This bracelet goes organically with the generous, but not plump, case, whose basic shape is in the style of the 1980s. The flanks are polished to a high gloss, while the top is brushed matte. The polished bezel, which looks like a stylized, knurled nut, contrasts nicely.

The massive case back, which, in addition to the positively engraved company logo and technical information, displays an elegant sailing ship, is worthy of mention. Inside, the case back is textured like a circuit board—completely unnecessary technically, but a sign of great care. The movement (diameter 26 mm) is mounted in a 42-millimeter case with the help of a mounting ring. This ring not only provides stability but helps protect against magnetic fields. The screening is completed by the soft iron dial and a dust cover, which is attached to the movement retaining ring by three small screws. This too isn't necessary, but ...

We come to the GMT function. Here the Overseas differs significantly from the other watches we have described. While the newly set local time appears on their watch dials as "main time," and home time plays only a secondary role, on the Overseas Dual Time it is exactly the opposite. If the crown is moved to the first detent, the hour hand on the small subdial can be moved to 6 o'clock, inclusive of the day-night indicator next to it. The date on the other hand is with the main time (in this case coupled with home time) and can be set quickly with the help of the pushbutton.

The manufacturer Caliber 1222C in the Overseas is shielded against magnetic fields by a soft iron inner case.

Mido
OCEAN STAR
datoday
Chronometer
OFFICIALLY CERTIFIED
SWISS
MADE
SAM 5
Mido
OCEAN STAR
26
Chronometer
powerwind
SWISS
MADE

In No Way "Retro"

They are the benchmarks in a world of watches characterized by all sorts of models: timepieces that are so attractive that they are built for decades virtually unchanged. One such model is the Mido Commander Ocean Star, which, for almost fifty years, has been prized by price-conscious watch lovers. Hopefully it will continue to be made for a long time.

The observer raises his eyebrows appreciatively: "Elegant retro watch," he says, praising the Mido Commander Ocean Star that gleams from under your cuff. He is right, and yet he is not. One can certainly credit the Mido with a certain elegance, but it is certainly no retro watch. Retro is understood to mean a modern product whose design is based on that of classic predecessors. But the Mido is no replica; it is the original model itself that first appeared under the name Ocean Star in 1959 and has been made unchanged ever since. This watch, which has stayed on the market for more than fifty years, is, at Mido, still an "important part of the collection." The Commander II, unveiled in August 2012, is not intended to change anything.

During the 1930s, Mido concentrated on simply designed watches, distinguished by being highly suitable for everyday use but with a touch of sportiness. In addition to having a robust movement and a sturdy case, the watches were waterproof, which was the exception at the time. Unveiled in 1934, the Mido Multifort achieved a milestone in the area of robustness. During tests it withstood a pressure of 13 bar (equivalent to the pressure at a depth of 130 meters of water). In the beginning, however, it was only guaranteed to a depth of 30 meters. What today is considered normal or even subpar with respect to precision-made cases and modern sealing materials was then pioneering.

A newly developed sealing system for the winder tube played a major part in the good test results. With respect to leakage, the entrance to the winding stem in the case is the critical spot on a wristwatch. We do not know if the solution to this problem was discovered while enjoying a bottle of wine, but it is at least likely. For the same material was chosen for sealing the watch as for sealing a wine bottle: cork. The crown seal was not cut straight from the bark of the cork oak, however. Instead the cork was finely ground, mixed with a bonding agent, compressed under high pressure, and finally impregnated with sealing grease. Thus processed, the material nestled close to the winder stem and kept water out of the case even when the crown was pulled.

Mido Commander Ocean Star

Reference: M 8429.4.C1.11

Movement: Self-winding, ETA Caliber 2836-2; diameter 25.6 mm; thickness 5.05 mm; 25 jewels; 28,800 A/h; certified chronometer (COSC).

Functions: Hours, minutes, central second hand, day of week, date.

Case: Stainless steel, diameter 37 mm, thickness 10.45 mm; hesalite glass; "Aquadura" cork crown seal; waterproof to 3 bar.

Bracelet: Stainless steel mesh (Milanaise) with fold-over clasp.

The ETA 2836 automatic movement has been made since 1972 and possesses outstanding reliability. As a tried and tested chronograph, it is worth having.

Model improvement made to measure. The (still) current Ocean Star (with day-date indicator) differs little from the classic from the 1960s.

It was not until a quarter of a century later that the child received a name. When the Ocean Star was launched in 1959, the cork seal that had been in use since the '30s was dubbed Aquadura. So it remains to this day, and the name is proudly marked on the back of the Ocean Star. This brings us to another special feature of the Swiss classic. The case and case back form one unit, which in expert circles is called "monocoque." With this the designers eliminated the (pressed- or screwed-on) case back as a potential source of danger. For the self-proclaimed "king of waterproof watches" it was another important step toward the "hermetically sealed watch."

In keeping with its special design, the movement is inserted from above into the watch, which is sealed with a pressed-on artificial glass. The inner edge of the glass ensures that the glass is seated firmly in the case and guarantees leakproofness. On the Mido, this component, also designated Réhaut, is printed with the minute markers and thus becomes part of the dial, especially as it also has twelve luminous markers. In the dark they show the positions of the massive raised hour indexes. In combination with the illuminated hands, they make it fairly easy to read the time in darkness. Purists complain that neither the minute nor second hands reach the minute markings.

In light, however, the current Ocean Star provides the wearer with a wealth of information. In addition to the time, it indicates the day of the week and date in a large framed window at the 3 o'clock position. This is updating the model in a measured fashion. The original model, driven by a Caliber AS 4117 movement made by the company's traditional supplier Adolf Schild, only indicated the time, while the model from the 1960s, shown in our historical photos, also provided the date. The current Ocean Star has an ETA automatic movement. Among watchmakers, the Caliber 2836 is considered very robust and reliable and—coming with a chronometer certificate—also very precise. It has been made since 1972 and may therefore rightly be seen, like the watch as a whole, as a classic.

Old (the upper watch in each picture) and new: the attention to detail with which the Ocean Star has been changed over the years is obvious from any angle.

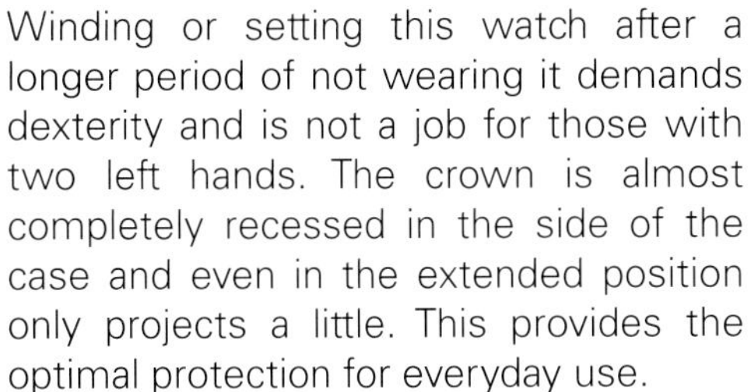

Winding or setting this watch after a longer period of not wearing it demands dexterity and is not a job for those with two left hands. The crown is almost completely recessed in the side of the case and even in the extended position only projects a little. This provides the optimal protection for everyday use.

The Milanaise bracelet is a characteristic of the Ocean Star. This is the maker's name for the close-meshed steel, which is very robust while also being very flexible. On the wrist the bracelet shows itself to be a cool skin flatterer and is suitably comfortable. The seamless adjustability of the bracelet length is also an advantage. But there is no light without shadow. The clasp is fiddly to operate and requires some practice. New buyers are advised to practice putting on the watch over a soft underlayer. Once it is on, it is secure. In recent years Mido has also added an additional safety catch. This small but practical update shows that Mido intends to retain this sprightly classic. *Ad multos annos.*

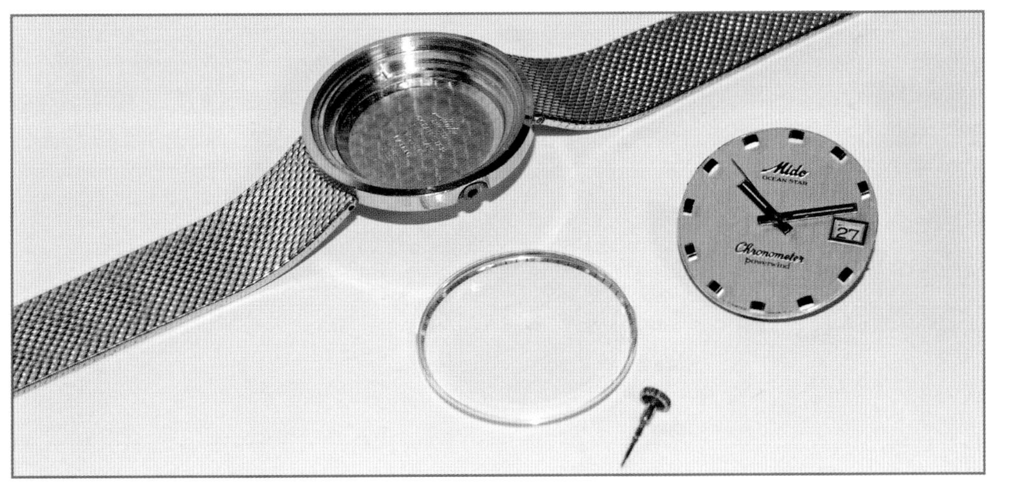

The one-piece case is also characterized as monocoque. The movement with dial and hands is installed from above. The glass is subsequently pressed into the case.

GIRARD-PERREGAUX
GP
TACHY
SWISS
MADE

Luxury Tickers for the Active

Carlos Dias, former owner of the manufacturer Roger Dubuis, coined the phrase at the SIHH 2004 Geneva Watch Fair: Sports Activity Watch or SAW. Large luxury watches that one could wear with a suit or on the beach were in vogue. Carlos Dias has since folded his sails, and no one speaks about the SAW today; however, these horological fireballs still enjoy a high level of interest.

The Easy Diver is a sporty chronograph with the look of a diver's watch. This is how the eccentric head of Roger Dubuis envisaged a Sports Activity Watch in 2004.

As Carlos Dias related at the SIHH 2004 Geneva Watch Fair, the idea came to him while on vacation: "It bothered me that I always had to take my watch off to go swimming," said the former owner of the manufacturer Roger Dubuis, whose products were conceived for horological aesthetes and not for active people. "I therefore promised myself that on my next vacation I would wear a Roger Dubuis that I never had to take off." But then a simple diver's watch seemed rather too banal to the native of Portugal: "We wanted to make a Porsche Cayenne of watches." To explain: a Porsche Cayenne looks like an off-road vehicle, weighs as much as a light truck, has the interior space of a mid-size station wagon and is as fast as a high-performance sports car. This combination of characteristics is subsumed under the term sport utility vehicle, or SUV for short.

With this approach it is clear that the new sports watch collection from the Geneva manufacturer could only be called SAW, or Sports Activity Watch. Dubuis started the sport watch business with three models, whose basic forms were taken from the well-known Sympathie, Golden Square and Much More lines. All of the watches were waterproof to 300 meters, with double anti-reflective coated sapphire glass and a rubber bracelet. In keeping with his motto "Just for Friends", Diaz limited all of its models: gold watches to 28 examples per model, steel and gold watches to 280 and the pure steel watches to 888 examples per model.

Roger Dubuis gave the square Aqua Mare (above) and the rectangular Sea More the look of diver's watches.

The top model among the Dubuis SAWs is the Easy Diver, a sport watch with the look of a diver's watch that is made as a self-winding three-hand watch, as a manual wind chronometer, or as a manual wind watch with Tourbillon regulator. The first two have an impressive diameter of 46 millimeters and the Tourbillon is even two millimeters wider. On all models, the crown is screwed on, as is, of course, the pushbutton on the chronograph. The Easy Diver is the only SAW that is a pure steel watch. The steel and gold variants offer perfect subtlety, for they cannot be immediately recognized as such, as they combine a steel case with a white gold rotating bezel.

From the Gold Square stems the rectangular sport watch model Aqua Mare with its rectangular case. When screwed in, the crown disappears between two massive guards. Also typical of the Dubuis sport watches is the galvanized dial, which is adorned with white gold markers. Superluminova in the two hands and in the indices ensure good readability at night. The Aqua Mare is driven by the company's Caliber RD 14 automatic movement, which also sees service in the rectangular Sea More. Apart from the shape of the case, it does not differ from the Aqua Mare and has, in defiance of the rectangle, a fixed bezel in the diver's watch style.

The Breitling Chronomat, which has long enjoyed classic status among watch fans, has a directional-rotating bezel. About 20 years after the unveiling of this successful watch, Breitling decided to update the classic. The trade likes to call this "restyling," and, at the beginning of the twenty-first century, it usually means increasing the diameter. This also applies to the Breitling, which, with a diameter of just under 44 millimeters, is now called the Chronomat Evolution. The Evolution did the pilot's chronometer good. The increase in size was achieved without altering the proportions and lines of the original model. That was in 1984, when Breitling celebrated its 100th year in business.

The model name Chronomat is derived from the words chronograph and automatic, for while the quartz watch was gaining importance, Breitling engineer Ernest Schneider retained the Valjoux 7750, a mechanical movement. That suited pilots—and continues to suit them today—because out of vital self-interest pilots usually prefer proven, reliable technology. The pilots of the Italian aerobatic team Frecce Tricolori were among the first wearers of the Chronomat, resulting in a special model restricted to 1,000 watches being dedicated to them. The Chronomat Evolution was also produced as a gold watch with leather strap, as a steel model, or a bicolor model in steel and yellow gold. The latter two are equipped with a massive bracelet. More on the Chronomat on Page 52.

The Breguet Type XX, a chronograph that was once conceived for the French navy, celebrated its 50th birthday in 2004. Of course, a new model was unveiled to mark the anniversary, the Type XXI. The Type XXI's case is larger than that of the Type XX, and, in Breguet fashion, the center section is fluted. The bezel rotates in two directions. Its face has also changed little. The subdial minute hand has been moved more to the center of the black rhodinised dial for improved readability, and the day-night indicator is at 3 o'clock and the date at 6 o'clock. And, as one would expect from a maker like Breguet, the Type XXI is also equipped with a new chronograph movement that has a flyback function. Despite all its new features, it can be said that the new Breguet has retained the sporting character of the old.

The Chronomat is a Breitling best-seller with origins back to the 1940s. Unveiled in 2004, the Evolution underwent a facelift and its diameter was increased to 33 millimeters.

Few watches in this market segment still have a diameter of 44 millimeters. Not even the manufacturer Girard-Perregaux, which, just one year after its launch, restyled the Laureato Evo 3 (illustration p. 36) chronograph in order to make it larger. The changes essentially affected the case and dial. Characteristic of the new silver-plated dial are the Clou-de-Paris décor in the center and the three subdials with their black rings: a 24-hour counter at 9 o'clock, an hour counter at 6, and a small seconds counter at 3 o'clock. Like the running second hand, the stop minute hand with the red arrow head comes from the center. In combination with the date indicator hand at 12 o'clock, the result is a rather unique watch dial. Even if the Evo 3 is called a sports watch, it is first of all a Girard-Perregaux, which must not be ashamed of its inner values. Therefore, the watch has a sapphire glass window in the case back, which is held in place by six screws.

Out of tradition the British feel an obligation to nautical time measurement, and, at the end of the eighteenth century, the so-called longitude problem was solved on the island. They proposed using the difference between the actual solar time and the local time on the Greenwich meridian during long sea voyages. It is a solution that survives to this day and is still used, when the GPS fails, for example. For this the seafarer needs the exact time plus a table with the time equation. The Longitude II by the British time-honored brand Arnold & Son offers both plus a little more.

On casual inspection the Longitude appears to be a chronograph, but the three controls on the right side of the case do not control stop-time measurement. The crown at 2 o'clock is used to adjust the rotating bezel beneath the glass with the longitude scale; the second crown at 4 o'clock is the actual winder, with which the time and date are changed. It also

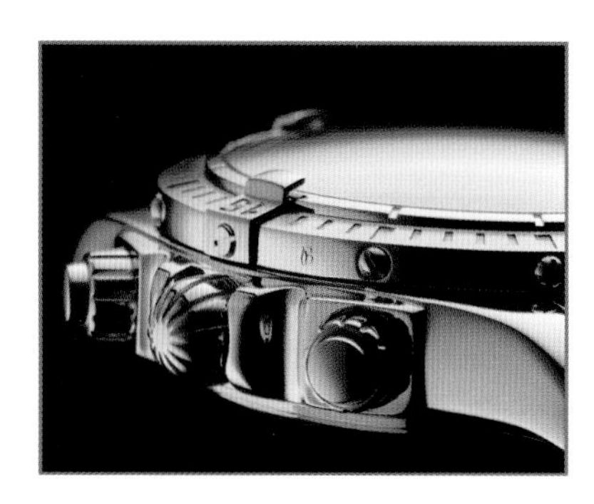

Breitling's attention to detail is shown by the small screw-mounted rider on the bezel.

controls the compass disc in the watch dial center. At 3 o'clock there is a simple button which, when pressed, opens a secret compartment in the case back in which the user will find the time equation tables. This watch has already withstood the harshest trials, for it went along on the expedition by Briton Bear Geylls. He led a group that crossed the North Atlantic along the Arctic Circle. Arnold & Son dedicated the special Arctic Model, with a yellow rotating bezel, to this expedition. Production was limited to 250 examples. The product from The British Masters is very British—except for the chronometer movement, which comes from Switzerland.

The same is true of the Graham brand, which also belongs to The British Masters. The new chronograph, first shown by the British in Basel, Switzerland, is called the Swordfish. It carries no sword or saw teeth, however, as its name might imply. Instead the watch seems to stare with googly eyes, for the chronograph's two auxiliary dials at 3 and 9 o'clock have a very unusual design. Clearly raised above the large sapphire glass are two small, slightly domed lenses, seated in their own bezels. They provide a clear view of the small second hand at 9 o'clock and the minute counter at 3 o'clock. Or the other way round, for there is also a Swordfish for wearing on the left arm with the crown and buttons on the left side of the case. Because the movement has to be turned 180 degrees for this, the auxiliary dial hands change their positions. The Graham easily won an unofficial competition for the largest wristwatch at the Basel fair with a claimed 46.2 millimeters.

Islands seem to inspire unusual shapes and enormous dimensions. Marco Mantovani, owner of the Italian watch brand Locman, lives on the Mediterranean island of Elba and there creates modern timepieces, some with mechanical movements, some with quartz ones. The latter is found in the Mare chronograph, which, with its 47-millimeter carbon case, makes a huge impression. The martial chrono does not burden the wrist, however, as the bezel, crown, and push-button are made of titanium, and the only burden is on the buyer's pocketbook. Locman asks about $1,000 for the Mare.

The mainstream follows none of the watches named here, which is very gratifying. On the other hand, the inclined reader may ask why one needs a Tourbillon, which can be worn diving, or a golden pilot's watch. We can no more answer these questions than we can explain why an off-road vehicle must be capable of 168 mph. There are people who simply want exactly that.

Facing page: *top left: Arnold & Son, bottom: front and side views of the Graham Swordfish, top right: Breguet Type XX, right center: Locman Mare.*

EAST
WEST
ARNOLD
ARCTIC
LONDON
FECIT

LOCMAN
ITALY
MARE

GRAHAM
Hours
Min
SWISS
MADE

CHOPARD
110
120
130
30
9
6
3

Sporty Chronographs

TACHYMETRE
JAEGER-LECOULTRE
SWISS MADE

Chronograph Fever

The man's only piece of jewelry is his watch, and his favorite toy—apart from his car—is his chronograph. Many feel sick if they are not wearing a luxury stopwatch on their wrist. No wonder. Chronograph fever is spreading. Here are several therapy suggestions from the year 2005.

No one really needs a chronograph. Unless, maybe, that someone is a man. He uses it to determine when his steak is medium rare, when his breakfast egg cooks exactly three and a half minutes, and how much time his favorite race driver gains on the leader per lap. These are entirely rational uses for the watch, and this is why the man spends long hours reading magazine articles and catalogs, and why he repeatedly creeps around watch store display windows. He is suffering from chronograph fever.

Not to worry, chronograph fever is curable. The therapy of choice differs greatly from patient to patient. The ticking medicines differ sharply in dosage form and content. Don't consult your doctor or pharmacist, just read the following lines.

Two extremely effective treatments for chronograph fever come from Jaeger-LeCoultre. They are the Maxter Compressor Chronograph and the Master Compressor Extreme World Chronograph (shown at left). Their contents: the Caliber 971 and 972. In both cases they are completely new chronograph movements with ratchet control and automatic winding. The Caliber 972 in the Extreme World Chronograph is also capable of indicating the time in different parts of the world and thus even helps those with homesickness. The sturdy packing in eye-catching steel cases promises a high level of effectiveness, and the 1,000-hour test by Jaeger-LeCoultre in 1992, long life.

The Chronoscope by Chronoswiss has demonstrated its broad appeal. In choosing the watch of the year for 2003, the majority of the readers of *ArmbandUhren* (*Watches*) and *Welt am Sonntag* (*World on Sunday*) chose this stopwatch as the ideal cure for chronograph fever. The concept and design of a timepiece with a push button integrated into the onion crown came from Gerd-Rüdiger Lang, an acknowledged chronograph specialist. Based on his own huge collection of chronographs, he continually researched new treatments for chronograph fever. In this case he retained the proven content (Enicar-based Caliber C.125) but gave it an attractive

Open: Chronoswiss Chronoscope with view of the control gear.

Clear: Zenith Grande Port Royal Concept with transparent dial.

Fast: TAG Heuer Caliber 360 concept chronograph.

packaging. An aperture in the dial reveals the control wheel and the associated lever that control the chronograph functions. The Chronoscope thus does justice to its Greek-derived named (chronos = time, skopein = to look).

Revealing the mechanism from the dial side has been a practice of Zenith for a long time. One promising product of this kind is the Grande Port Royal Concept. It not only helps cure chronograph fever, but thanks to its extremely masculine packaging, it also contains a considerable dose of testosterone. The rectangular case of matte-grey titanium measures 36 x 51 millimeters and accommodates a dial of TR 90, a transparent material that comes from spaceflight technology. With its folding clasp, a strap made of Kevlar and carbon ensures a snug fit on a large wrist. Zenith did not experiment when it came to the contents, however. In the new Port Royal ticks the proven El Primero, which measures time with an accuracy of 1/10 of a second.

Even more precise is the Caliber 360 by sister company TAG Heuer. In their archive, the watch technicians from La-Chaux-de-Fonds discovered a formula for a movement with which one can mechanically stop to one-hundredth of a second. The size of the micro timer, first shown in 1916, no longer appeared contemporary to TAG Heuer, and so they reduced the chronograph movement from pocket watch to wristwatch format. The TAG Heuer concept, which combines a movement for telling time and a movement for stop-time measurements, is unique.

When ill, tough men need strong medicine. Audemars Piguet knew that for a long time and relieved patients quaking with chronograph fever with various versions of the Royal Oak Offshore. With its prominent, octagonal bezel, the massive case of brushed stainless steel alone signals strength. The special Alinghi Polaris model, for which the watchmakers in Le Brassus developed a special movement, helps against homesickness. The Caliber 2326/2878 not only has a chronograph with flyback function, but also, when needed, it has a a window at 12 o'clock that displays the last 60 seconds before the start of a sailing regatta. With all these advantages, one cannot fail to mention the side effects: the Royal Oak Offshore can make one addicted and in the form of the Alinghi Polaris can also cause seasickness.

Strong: Audemars Piguet Royal Oak Offshore Alinghi Polaris with regatta function.

That is unlikely with the Omega De Ville Rattrapante. It is made primarily for patients who want exclusive ingredients in tasteful cases. Omega fulfills this wish by packing the chronometer-checked chronograph movement 3612 with a drag indicator function and co-axial movement in a simple stainless steel case that echoes the style of the 1950s. Omega does, however, give its customers a little optical extravagance through a unique dial design. The dial cutout for

Elegant: Omega De Ville Rattrapante.

Versatile : Glashütte Original Sport Evolution.

the date indicator is in an unusual place between 11 and 12 o'clock, while the permanent seconds hand stands out because of its unusual scale. Instead of the usual second dial, here one sees two concentrically-arranged semicircles. The inner scale extends from 0 to 30, the outer from 30 to 60. The two unevenly long ends of the second hand provide an unmistakable indication of the running time. The Omega does not begin lowering the fever immediately, but its effects are long-lasting.

Whenever chronograph fever strikes—whether during sports or in leisure time, whether in tough business meetings or while attending the theater—the Sport Evolution Chronograph by Glashütte Original is there to help. The Saxons have given the chronograph movement based on the Caliber 39 a contemporary cover. The chronograph's round case is clearly more striking than its dodecagonal predecessor and it may cause more patients who prefer medicine from German manufacturers to reach for a stopwatch from the Müglitz Valley. All the same, the watch retains an individual identity thanks to an attractive dial.

Some need originals to stay healthy, others get by very well with generics, especially if they are based on the dependable products by ETA. Oris of Holstein has a great deal of experience and a long tradition in the finishing of basic movements for other manufacturers. Visually, the remake of the Chronoris, a successful model from the year 1970, is so unique that even experienced fever patients guess that it is an independent development. Thanks to the reduction of the secondary dials to a minute counter at 12 o'clock, it requires a second glance to realize that inside the chronograph is the reliable panacea ETA 7750, known to many of the afflicted simply as Valjoux. Here something good from the past is combined with modern additions, like a new crown quick release fastener. That is rarely a poor choice.

Downsized: Oris Chronoris.

Paul Picot also reached back to the same original product as Oris. Only the developer of the company from Le Noirmont in the Swiss Jura added something to the ETA 7750. The movement was extensively revised, making possible a unique three-dimensional dial display. The indication of hours and minutes takes pride of place. Beneath this raised dial circle are the hands for the small permanent seconds at 9 o'clock and the stop time minute counter at 3 o'clock. Like the previously mentioned Omega, the two hands are of different lengths and thus brush their semicircular scales. But because the hand axes and the "unemployed" halves of the hands are hidden under the main dial, the whole thing has the look of a retrograde display. Word of this placebo effect will surely spread quickly among those plagued by fever.

Original: Paul Picot.

This subtle approach to chronograph fever is not shared by the French maker Bell & Ross, nor by the Swiss brand Formex 4 Speed. Both pack their active ingredients combined with a good portion of adrenaline in huge square steel cases—regular hospital packing that suggests immediate and lasting effect. The Instrument BR01-94 by Bell & Ross, with its crown and push button on the left side and an easily-readable dial, looks at least halfway conventional and therefore is easy on the circulatory system. On the other hand the Formex drives the pulse to unanticipated heights, for on the dial there is pure chaos: many colors, many shapes, many indications. In addition to the chronograph function, there is also a GMT indicator, plus a rotating inner bezel with no less than three scales. The Formex is the ideal watch for someone who will recover from his fever according to the motto "the more, the better".

Adrenaline pumping: Bell & Ross BR01-94 (top). Wild: Formex 4 Speed (bottom) with overhead pushbuttons.

Chopard shows that angular cases do not always have to be flashy. With its elegant exterior, the rectangular Chopard Tycoon looks so serious, so aristocratic, that one cannot attribute any negative side-effects to it. It is therefore entirely possible that people will turn to Chopard who do not even suffer from chronograph fever.

On the other hand, the Temerario is the best medicine against the "febris italiensis." The Eberhard Company of Tessin here used a recipe that previously

Aristocratic: Chopard Tycoon.

Moonstruck: Longine Master Collection.

proved itself in the round Chrono 4: the chronograph movement's four subdials are arranged in a row—lengthwise on the Temerario. The movement was simply rotated 90 degrees counterclockwise, causing the pushbutton and crown to be atop the elegantly-finished case. This gives the watch a very special look and reduces the fever in a very comfortable way. They are similar to the chronographs from the Longines Master Collection, except that they express Swiss respectability instead of Italian charm. This definitely will not cause the wearer to yawn; instead it creates the comfortable feeling of not having missed the mark. Longines even assists the moonstruck, for the top model of the Master Collection has a phase of the moon indicator in addition to a calendar.

Eberhard & Co. Tremerario with all subdials in a row.

The moon also plays an important role at Fortis, for the Swiss maker has been testing its products in space for more than ten years. Pictures of smiling cosmonauts from the ISS international space station demonstrate impressively that the Fortis chronographs work reliably even in space. Because the crew of the tenth ISS mission landed safely, even down-to-earth watch fans can now buy a space-tested chronograph: the Fortis B 42 Official Cosmonaut's Chronograph in titanium. This watch with the logo of the Russian space agency FSA and a picture of the ISS station on the case back is, however, limited to 500 examples; consequently, not all of those suffering from chronograph fever can be treated. Fortunately the previously described fever-reduction medicines are very effective. Unfortunately, they are not available by prescription.

Fit to fly: Limited Fortis B-42 Official Cosmonauts Chronograph.

ZENITH
EL PRIMERO
36 000 VpH
CHRONOMETER
SWISS
MADE
TACHYMETER
SEIKO
Ananta
AUTOMATIC
IWC
SCHAFFHAUSEN
YACHT CLUB
SWISS MADE

Manufacturer Movements

Chronograph lovers who are prepared to make a four-digit investment encounter the same movement in many brands: the ETA-Caliber 7750, called Valjoux, because of its place of origin. This price segment definitely offers room for individuality in the field of watch movement technology.

A chronograph is first of all nothing more than a wristwatch with an integrated stopwatch. So simple the definition, and yet, so limited seems the choice of the movement: the Caliber ETA 7750 "Valjoux"—named after Vallée-de-Joux, the Swiss watch valley and built by the millions since 1973—dominates the mechanical chronograph market. Such watches with stainless steel cases command prices in the mid to high four-digit dollar range.

Now, the Valjoux is undoubtedly a proven, reliable motor, but ambitious chronograph lovers want something special, especially if they are prepared to spend more than

Limited to 2,000 watches, the limited special edition of the Chronomat B01 has a black dial and red-painted subdial hands.

$3,000. Alternatives to this mainstream movement are certainly not legion, but they can definitely be found. It has since become good form for the renowned Swiss makers, who long built on the 7750, to also offer their own chronograph movements. In 2007 IWC Schaffhausen first presented its Caliber 89360 in the Da Vinci. Breitling followed one year later with its own Caliber B01. Finally, in 2010, Tag Heuer brought out the Caliber 1887, which was based on a Seiko movement. Because of its sheer size, the Japanese watch giant had always been very self-sufficient and in the '60s it was in a race against the Swiss to develop the first self-winding chronograph movement, which Zenith won by a narrow margin in 1969. The result was an illustrious field with widely differing characters and entry prices, which ranged from just under $4,000 to $14,000.

Breitling Chronomat B01

Flyer with Window

The Breitling brand stands for watches with instrumentation like few others and has devoted itself primarily to aviation. It has sponsored owners of historic aircraft, as well as entire air displays, in particular the legendary Reno Air Races. Prominent pilots like John Travolta are among the celebrities who spread the company's message in full-page advertisements. That creates an image. Among pilots, Breitling is considered a typical pilot's watch. On the other hand, some watch aficionados revile Breitling as a high-price ETA box maker. In the opinion of Breitling owners and management, the latter was a disgrace that had to be taken care of, and so in the summer of 2004, under conditions of extreme secrecy, the company began development of a chronograph movement that would be "100 percent Breitling." As the watchmaker sold more than half a million timepieces in a year, the movement had to be able to be produced on an industrial scale, with acceptable costs and no reduction in quality. Finally, precision was part of the self-image of a brand that sent every watch to the COSC for chronometer testing.

The glass case back was also reserved for the special edition. The standard Chronomat has a steel case back.

With the help of outside specialists, the development process literally flew by: 2006 saw the first design drawings, the first prototypes, and the first movements, which underwent COSC testing, followed by industrialization and a three-point landing. At BaselWorld 2009, Breitling unveiled the first chronograph with its own Caliber B01 on the mark's 125th birthday.

For those who like technology: the Caliber B01 is a column wheel chronograph with automatic winding and vertical coupling between motor and stop mechanism. Here the stopwatch's hour counter turns continually, while in classic chronographs, like the ETA 7750 Valjoux, it skips. Other Breitling specialties are the patented auto-centering mechanism and regulation of the watch by a rotor in the balance wheel.

The buyer of a B01 chronograph normally sees none of this, as it has a steel case back. Good for those then who obtained a model from the special series, limited to 2,000 watches, for a steep premium of $1,400. Not only did they give it an extremely sporty look with black dial and red stop hands, but a viewing aperture in the case back screwed to the case.

Crown and pushbutton are screw-down, and even with the glass case back, it claims to be waterproof to 20 bar (200 meters), while the unlimited series watch with steel case back is good to 50 bar. This is a watch like a safe. The steel bracelet with folding clasp not only holds the watch securely and comfortably on the wrist, but also acts like a balance weight. With a leather or rubber strap, the Chronomat would probably be rather top-heavy. Several areas that come up for criticism are the highly polished surfaces on exposed areas, which tend to scratch, and the bracelet. It does not quite fit the hard-core image of an otherwise all-round successful sports watch.

IWC Portuguese Yacht Club:

Under Full Sail

The story of the IWC Portuguese watches goes back to two Portuguese watch salesmen named Rodrigues and Teixeira, who, in the 1930s, went to Schaffhausen looking for a timepiece with the qualities of a ship's chronometer that could be worn on the arm. Not at all in keeping with the style of the times, it was to be a large and especially accurate watch.

The IWC Portuguese Yacht Club is linked to the water in many respects. It is powered by an 89360 chronograph movement made by IWC.

Crown guards and large pushbuttons give the Yacht Club sportiness.

That remains IWC's objective today: clarity, size, precision, and a sophisticated mechanism make an almost perfect symbiosis. This is also true of the new Portuguese Yacht Club.

The Yacht Club was initially a robust sports watch with three hands, one of the most successful IWC timepieces of the past century. As part of a reorganization of the model families under the aegis of Georges Kern, the new Yacht Club now finds itself in the Portuguese family. And not only the nomenclature has changed. The new edition also brings a measure of sportiness to the otherwise very elegant world of the Portuguese.

With the IWC manufacturer Caliber 89360, it has a modern and robust chronograph movement on board. The movement had its premiere in 2007 in the then-new Da Vinci family. This gear-controlled manufacturer movement shines with several very special design features. Of particular note is the stop time indication, which in terms of user friendliness is a departure from classical models. The stopped hours and minutes can be read like a second time of day on a subdial at the 12 o'clock position. The Caliber 89360 also shines with a new dual catch winding system and a flyback function: pressing the button at 4 o'clock allows the stop time to be set at zero and immediately begin running again, without halting the stop mechanism.

The watch is waterproof to 6 bar and is the only Portuguese with illuminated hands and markers. It is thus readable in darkness, which is very welcome in everyday use. With massive flanking protection for the crown, not only is the IWC robust and functional, bit its exterior also looks very sleek and elegant, not unlike a sailing yacht. The rubber strap with integral folding clasp fits harmoniously into the total picture and the quality of workmanship is evidence of great care, but one can only expect this with a starting price of close to $14,000.

Seiko Ananta Chronograph:
Swords to Watch Cases

The Seiko Ananta Chronograph occupies a special position in this quintet in several respects. The Seiko is the only watch in the field that is not from Switzerland; it is also the most inexpensive watch, which should not be confused with cheap.

Finally, it shines because of its special case design. The case back and the strap lugs are milled from a single piece of steel, which makes the parabolic curve possible. The case center section is attached to this base element by six screws.

The designers combined convex (center section) and concave (strap lugs) shapes. This creates visual intensity on the one hand, and on the other, allows the watch

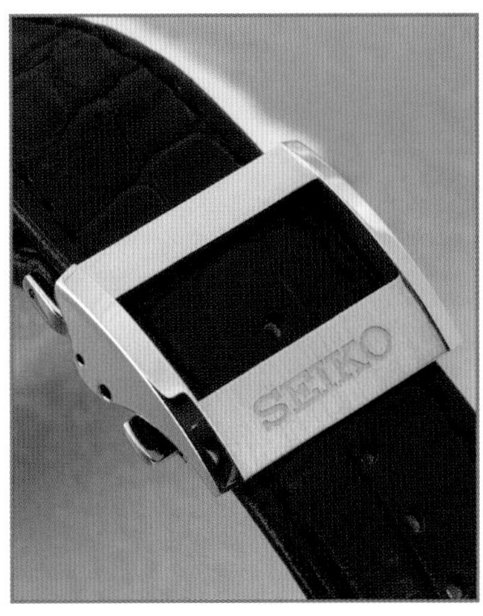

The Seiko Ananta is the price-performance monster among the five manufacturer chronographs. It costs about $3,700..

to appear very powerful, which is entirely deliberate. Finally, the katana sword, a weapon first built 800 years ago and a legend in Japan to the present day, serves as inspiration. Because of its ostentatiously powerful design, the 46-mm monster is enormously attractive to younger watch fans. They should, however, have powerful wrists, for otherwise the Ananta tends to slip on the arm because of its wide and largely flat supporting surface. In any case, the reptile leather strap, whose proportions fit well into the overall picture, should always be closed halfway tightly.

The Seiko chronograph is also anything but reserved, technically. The manufacturer Caliber 8R28 has a control wheel that controls the chronograph functions start, stop, and zeroing. Each of the three chronograph counting wheels has a vertical clutch, so that all of the chronograph hands start when the easy-to-use buttons are pushed. The vertical clutch does not employ gear wheels to transmit force, but instead uses a friction mechanism. Seiko prides itself in having become the first watch maker to use the vertical clutch and column wheel together in a chronograph, the Caliber 6139, in 1969. The special features cannot be seen through a window in the case back, but a look through the porthole is enough to convince the observer of the movement's flawless workmanship, which can, in fact, be said of the entire watch. Only the operation of the crown, which unfortunately is not screw-down, allows minor doubts to arise, for in hand setting mode it gives the impression of literally being uncontrolled. This can only slightly sully the positive overall impression of this price-performance king.

The Seiko's powerful optics are attractive to younger watch fans. Unfortunately, the control gear cannot be seen through the window in the case back.

TAG Heuer Carrera Caliber 1887:

Homage to the Founder

Everyday functionality has always been a priority at TAG Heuer. "The movement of our chronographs stands out because of its great simplicity," declared company founder Edouard Heuer in his patent application for a novel chronograph movement in April 1882. At that point Heuer had been making chronographs for two years. Heuer's declared goal was

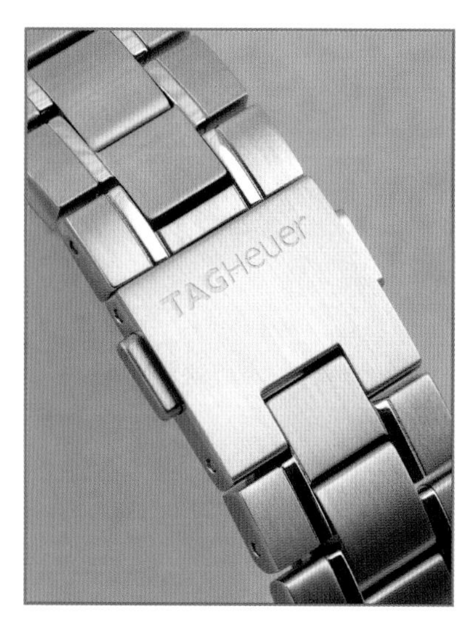

The Carrera Caliber 1887 looks functional and tidy in the 41-millimeter case.

to democratize the stopwatch. Until then, chronographs had been very expensive because of their design with control wheel and clutch, and were beyond the reach of most people. Simplification was called for.

The Caliber 1887 movement is based on a Seiko design but it is largely produced by TAG Heuer itself in Switzerland.

Edouard Heuer's efforts resulted in a patent, which was issued on 3 May 1887. The small part that revolutionized the chronograph world was called an oscillating pinion. It was a vertically-swiveling shaft with two pinions. One engaged the movement, or the motor. When the stop button was pressed, the shaft swiveled enough to cause the second pinion to engage the gears of the chronograph mechanism. The stop second hand was propelled by direct traction.

Obviously the first chronograph movement produced in series by TAG Heuer in-house used an oscillating pinion as chronograph clutch. The name of the watch was an homage to the patent of Edouard Heuer and was therefore simply called the Caliber 1887. This caliber was not an entirely new design, however. Instead, it was based on the Caliber TC 78 by Seiko Instruments, whose exclusive manufacturing rights TAG Heuer had acquired in 2006. Even though the basic design was Japanese, the Caliber 1887 may call itself "Swiss made". Ultimately the movement was redesigned by TAG Heuer in order to permit the use of an escapement part—consisting of escape wheel, rotor, balance wheel and hairspring—from the Swiss components maker Nivarox-FAR. TAG Heuer also installed a ball bearing mounted winding rotor. Not surprisingly, the CAB 1887 found its first home in a Carrera. This name has been synonymous with (TAG) Heuer sports chronographs since 1964.

The variant illustrated here lacks the tachometer scale engraved in the bezel—so loved by racers. Because of its classic design, the chronograph, with its steel band and reserved look, also goes perfectly with a suit. A case thickness of 16 millimeters, however, does not go with delicate and sometimes tight cuffs. The fact is: with a starting price of about $5,500, this timekeeper is one of the most reasonable Swiss-made chronographs not driven by the universal Caliber ETA Valjoux 7750.

Zenith El Primero Capitain:
The Gentleman Chronograph

How gratifying: After ex-CEO Thierry Nataf was responsible for years of bizarre design ideas and excessive prices, under the leadership of Jean-Frédérique Dufour, Zenith returned to its traditional values—tradition, sophisticated watchmaking, and a realistic price-performance relationship. Representative of this development is the El Primero Capitain shown here. According to information from the European representatives of the LVMH Group (Louis Vuitton, Moët, Hennessy), this model is enormously popular in Germany.

Here, tradition-conscious Zenith fans will find what they long missed: a slim, sleek and elegant chronograph. With a diameter of 42 millimeters, the size of the case is contemporary. The Caliber 400B movement, whose basic design dates from the '60s, has not kept pace, however. Zenith unveiled the Caliber 400 in 1969 and proudly called it El Primero. The company ultimately beat out its competitors—Seiko of Japan and the consortium of Heuer, Breitling, and Hamilton Büren by a hair with this movement. At that time men's wristwatches had a diameter somewhere between 36 and 38 millimeters and the movement was made correspondingly small, ending up at 30 millimeters. It therefore required a movement retaining ring to make up the difference in size between the movement and case.

None of this is apparent to the observer, as it is elegantly hidden by a case back with numerous engravings. Through the window can be seen the movement, whose architecture has lost none of its

beauty over the years. And, technically, the high-speed oscillator remains something special. The balance amplitude of 36,000 half-oscillations per hour, which Zenith proudly advertises on the watch dial, makes the chronographs accurate to one-tenth of a second.

Purists may criticize the fact that the lack of proportion between the movement and case sizes causes the subdials to be forced into the center. The reply is that Zenith has found a quite appealing solution, modifying the date wheel and moving the date window to the 6 o'clock position. With its applied steel indexes and its symmetry about the longitudinal axis, the dial has a very balanced look. Weighing about 80 grams, the El Primero Capitain is comfortably light and therefore manages even with a fine folding clasp on the crocodile leather strap—a true gentleman among chronographs.

With the El Primero Capitain, Zenith makes a chronograph that can easily be worn with a dark suit.

The design of the El Primero high-speed oscillator has its origins in the 1960s, but it is still a feast for the eyes in the twenty-first century.

TAG HEUER
Grand
CARRER
CALIBRE
RS
AUTOMATIC
OFFICIALLY CERTIFIED
CHRONOMETER
TACHYMETRE
8

TAG Heuer and sports

Few other watch brands stand for sports like TAG Heuer. The Swiss are traditionally associated with sports, especially with motor sports. The sophisticated, luxury brand has long been a good seller, but when one speaks about chronographs, one also always speaks of TAG Heuer.

Actor Steve McQueen, though no longer with us, helped make the square Monaco chronograph world famous.

From the start, TAG Heuer was associated with sports, especially sports time measurement. Equestrian sports, light athletics, ski racing—since the company was founded in 1860, making high-level performances in sports measurable has been one of the core competencies of the Swiss watchmaker, seen by the general public as a specialist in chronographs and sports watches. But Heuer—and later TAG Heuer—was most closely linked to motor sports.

This is an image that the Swiss subsequently cultivated in their communications, and their sales success also had much to do with clever marketing. The company thus positioned itself as a sporting brand—especially after the company was renamed TAG Heuer, which launched campaigns that caused a furor in the advertising scene. The "Success, It's a Mind Game" campaign (1995), whose protagonists were sports figures celebrating success under extreme pressure, has become almost legendary. A typical advertising motif: a swimmer who is pursued by a shark but is not eaten. Mental strength and the persistent striving for perfection are the key messages of this campaign, which was transported into the early 1990s with the "Don't Crack under Pressure" tagline. Racing legends Ayrton Senna and Michael Schumacher played central roles in the advertising campaigns.

The Swiss have always used race car drivers as advertising vehicles—also called market ambassadors. Jack W. Heuer, great-grandson of founder Edouard Heuer, began sponsoring racing sports in 1971, supporting the Scuderia Ferrari as official timekeeper and equipping the drivers with Heuer chronographs. Of course, these drivers also wore the Heuer

The Chronograph Mikrogirder crosses the boundaries of classical watchmaking and mechanically measures time with an accuracy of one ten-thousandth of a second.

Further proof of TAG Heuer's competence is the Mikrotourbillon S, whose movement can be seen here on a designer's computer screen.

logo on their racing overalls, especially Swiss drivers like Clay Regazzoni and Jo Siffert, who drove in Steve McQueen's racing epic *Le Mans*. In this way, the Swiss watchmaker got on film, earning it a pinch of fame. This was an entirely welcome development that continued to be pursued. The company's "What are you made of?" advertising campaign not only featured Formula 1 champion Lewis Hamilton, but film stars like Uma Thurman and Leonardo Di Caprio as well.

To attribute TAG Heuer's success solely to good marketing and smart advertising would, however, clearly fall short of the truth. Traditionally the Swiss have possessed extensive technical know-how, which is emphatically underscored by a look at the brand's technical history. In 1887, Heuer invented the so-called oscillating pinion, a brilliantly simple chronograph clutch, which was also patented. In 1916, the Mikrograph followed, designed by Charles-August Heuer. It oscillated at 360,000 half-oscillations per hour (A/h) and thus was able to measure time with a precision of hundredths of a second. The Carrera Caliber 360 unveiled in 2005 was capable of the same thing. The chronograph is equipped with two movements, which, though connected to each other, operate independently. The time of day is indicated by a chronometer-precise automatic movement that ticks with a frequency of 28,000 A/h, while stop time is measured by a movement which, like its forebear from the year 1916, measures accurately to hundredths of a second but is significantly smaller. The chronograph movement can be used to record time up to 100 minutes, and only then must the time measurer be wound by hand. One technical refinement: both movements can be wound with the same crown.

This so-called "dual chain" principal with separate movements also follows the latest highlight from La-Chaux-de-Fonds, the MikrotourbillonS. As the name suggests: both time indication and measuring are controlled by a tourbillon, with the chronograph tourbillon oscillating at an unbelievable 360,000 A/h or 50 Hertz. That is a world record. With this watch, TAG Heuer and its ambitious head of development Guy Sémon combine traditional watchmaking with highly technical innovation.

In the TAG Heuer range, such demonstrations of competence are placed in the "concept watch" category, which also includes the Monaco 360 LS—with the same combination of movements as the Carrera Caliber 360—the Monaco V4 with gearwheel drive, and the SLR Chronograph. The latter is the result of a cooperative effort with the Daimler AG and has a minor technical refinement in the shape of a chronograph pushbutton, which is operated from above. The first version of this model is intended solely for buyers of the Mercedes SLR high-performance sports car and has the vehicle's chassis number engraved in the case back. Meanwhile, TAG Heuer has produced other SLR models that are generally available.

Customers with an affinity for automobiles also favor the Monaco, Carrera, and Grand Carrera models, whose forms resemble those of older, successful Heuer classics. These are permanent elements of the TAG Heuer collection that are regularly augmented by interpretations of older successful models. In 2003, for example, the company released a replica of the Autavia (automobile/aviation) chronograph, which was so called because it cut a good figure in an automobile or an aircraft cockpit. The Autavia was powered by the new Caliber 11, a chronograph movement in which, following an old tradition, the crown is positioned at 9 o'clock. This movement is a new design by Dubois-Dépraz, based on the ETA 2894, which in certain respects is already a tradition. Dubois-Dépraz had previously taken part in the designing of the original Caliber 11, which was unveiled in 1969. It is not surprising, therefore, that TAG Heuer presented another watch with the Caliber 11 in its anniversary year: the sporty-elegant Model Silverstone, once the favorite of Clay Regazzoni.

As a general rule, mechanical TAG Heuer chronographs and watches are driven by modified ETA production movements. Happily for consumers, the TAG Heuer Prestige is still offered at halfway reasonable prices. Automatic watches with steel cases and bracelets can be had for under $2,700, a chronograph with modified ETA 7750 (Valjoux) costs about $4,100. Fans of the brand have to reach deeper into their pockets if they want to purchase a chronograph with the added designation Caliber 36. Inside these timepieces are the more complex and thus expensive El Primero chronograph movements made by sister company Zenith. Finally, at the BaselWorld Fair in 2010, the Swiss unveiled a chronograph with its own movement, developed in cooperation with Seiko. It answers to the name Caliber 1887 and thus recalls the mentioned oscillating pinion, which is also used in the new movement.

With this large range of mechanical watches, it is easy to forget that TAG Heuer has traditionally always had great technical expertise in quartz watches. In 1966, Jack W. Heuer, grandson of the company founder and from 1958 operationally responsible in the company for twenty years, built the first compact electronic timepiece with accuracy of one thousandth of a second: the patented Mikrotimer (later Microtimer) of the 1960s served as pattern and name-giver for an avant-garde digital wrist chronograph, which, as before, was favored by Formula 1 fans. And the youthful entry-level Formula 1 collection, promoted especially by race car driver Kimi Räikkönen, was controlled exclusively by a quartz crystal.

This exploded view shows the complexity of the Caliber 1887 chronograph movement.

The Monaco 69 hybrid chronograph combines the two worlds of TAG Heuer. It has a mechanical movement for displaying time and a quartz-controlled digital chronograph. A complex reverso case allows the wearer to choose displays.

GP
GIRARD-PERREGAUX
MONTE-CARLO 197
TACHYMETRE 600Km/h
SWISS MADE
TACHYMETER
km/h
Glashütte
ORIGINAL
GLASHÜTTE I/SA
MADE IN
GERMANY

Retro Classics

In creating the Monte-Carlo 1973 by Girard-Perregaux and the Glashütte Original Senator Chronograph XL, the designers looked in the rearview mirror. The result were stylish classics with very different characters, whose technology and finish are in no way outdated.

In the world of fashion, the 1970s are generally considered somewhat difficult. One thinks of bell-bottoms, patterned shirts with huge collars. Today one would only wear such clothes to a carnival or appropriate 1970s parties. It is much different when it comes to cars and watches that interpret the style of that decade. As we see and hear, they continue to be popular, especially among the generation now in its late 1940s and early 1950s. The reason? We can only surmise: while we actually wore bell-bottoms and polyester shirts, back then we had to get by with simple timepieces and means of transportation, because we could not afford the finer things. Now things are different. In the so-called best years of his life, a man is not only willing but also able to lay out a four figure sum to take stylistic revenge.

The Monte-Carlo 1973 by Girard-Perregaux would be ideally suited in a number of respects. This watch brings together fine watchmaking and motor sports enthusiasm. It is a combination that goes back to the late GP chief designer Luigi "Gino" Macaluso. Macaluso, himself a successful rally driver, began his second career as watchmaker by dedicating a model limited to 250 watches to winners of the Monte Carlo Rally. It was always important to him to ensure that the style and design of the watch corresponded to the epoch and the victorious car.

He accomplished this in the Monte-Carlo 1973, which paid homage to the Renault Alpine 110 and its driver, Jean-Claude Andruet. In doing so he refused to lower the quality; otherwise the case would have been made of plastic, like the body of the French rally car. The Swiss refused to do this and instead designed a stainless steel case that was finely made down to the details and was capped by a moulded sapphire glass. The edge of the glass is adorned by an engraved tachometer scale, which makes speed measurements possible, and the blue-painted stop second hand and the blue numbers in the black subdial recall the traditional blue paint of the Renault Alpine.

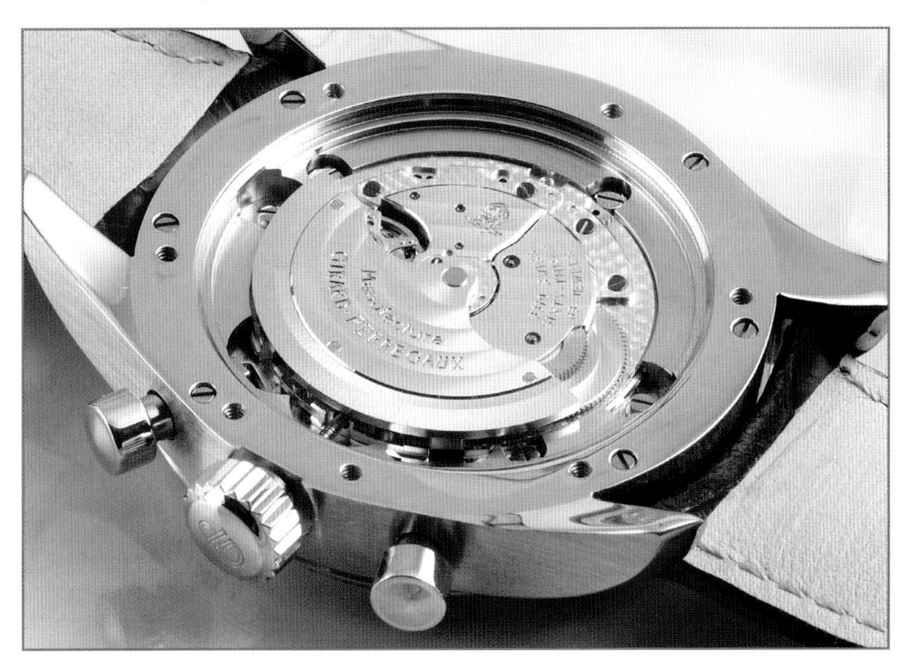

With a diameter of 29.4 millimeters, the manufacturer Caliber 337A has plenty of room in the voluptuous steel case.

The Senator Chronograph XL scores with solid elegance. The broad movement retaining ring is normally hidden by the case back.

The silver dial and the polished, illuminated hands have a first-class finish and are beautiful to look at. But the wearer must suffer a little because of this beauty, as the low contrast between dial and hands result in less than ideal readability. Normally the beauty of the movement is hidden from the observer—beneath a steel case back, in keeping with the era. This offers ample surface for engraving on the theme of the watch. The official rally shield with the number 18 is to be seen here along with the vehicle's type designation and the names of driver Jean-Claude Andruet and his co-driver "Biche" (Michèle Espinosi-Petit). Beneath this cover is a fine motor. It is based on the automatic Caliber 3300, which, with the addition of a module, became the chronograph Caliber 337A, which also has a flyback function. The drive has no need to complain about lack of space. On the contrary, a massive movement retaining ring makes up for the size difference between the 29-millimeter movement and the 40-millimeter case. Functionality does not suffer, and crown and pushbuttons function flawlessly.

Of these two watches, the Girard-Perregaux clearly takes the place of a sports car, whereas the Senator Chronograph XL by Glashütte Original embodies the qualities of a dignified limousine—a Mercedes perhaps? In any event, in designing the case and dial, the Saxons looked somewhat further back than the Swiss, with styling elements from the 1950s and 1960s. Nevertheless, the Senator definitely does not look old-fashioned, which it owes in no small part to the balanced nature of its dial. The dial is pleasantly calm, because a chronograph hour counter and date indicator were dispensed with. The Saxons very reservedly placed the tachometer scale on the inner glass bezel, the Réhaut. With respect to contrast and readability, the Senator is on the same level as the Monte-Carlo, and the hands and dial are tone in tone. Glashütte Original does, however, offer the Senator chronographs with a black dial, which those who favor good readability may appreciate.

Like the GP, the Glashütte Original is a module chronograph—in this case on the basis of the Caliber 39. From the mid-1990s this automatic movement literally got the re-privatized Glashütte watch company going again and even today is still an integral part of the watch program, which by now includes more than a dozen different watchmaking plants. Over the years the design of the high-torque drive has not only proved extremely robust but remains pretty to look at. Watch lovers value the traditional Glashütte features like three-quarter plate, winding gears with Glashütte sunburst, or gooseneck precision adjustment. The skeletonized rotor with 21-karat-gold gyrating mass is both an attractive and effective energy provider.

The precision-made movement is packed in a stainless steel case with a diameter of 44 millimeters, which answers the modern watch buyer's desire for large watches. But because the movement comes from a time when smaller watches were in demand, a movement retaining ring is required. In keeping with the company's own expectations, it is finely

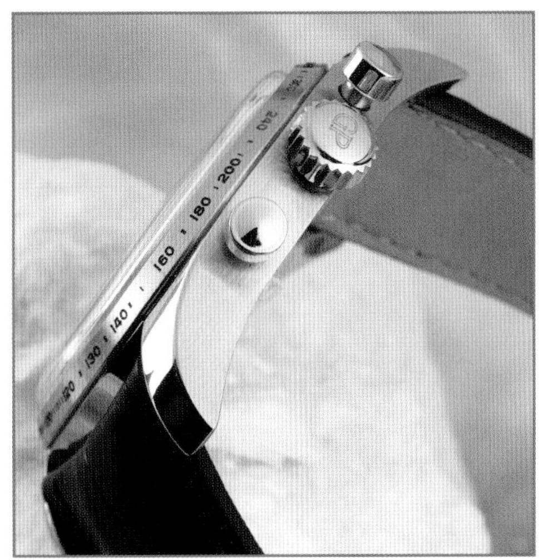

A livelier dial and a bezel with engraved tachometer scale demonstrate the Monte-Carlo's sport nature. Girard-Perregaux has since removed it from its catalogue.

made and in any event cannot be seen during normal operation, being hidden by the steel case back. The viewing window of sapphire glass is precisely large enough to allow the movement to be seen and appreciated. The case's proportions are generally very balanced, while design elements like the elongated pushbuttons and the stepped bezel give the Senator a very slim appearance.

Ultimately the Senator is just as attractive to the age and target group mentioned at the beginning as the Monte-Carlo, even if it does not exactly awaken youthful memories. In terms of price, both watches are in the upper four-digit range, which unfortunately must be seen as reasonable for a manufacturer chronograph these days. The Monte-Carlo is no longer in the current catalogue, however, and can only be obtained as a new item with a great deal of luck.

Glashütte Original Senator Chronograph XL

Movement: Self-winding; Caliber Glashütte Original 39-34; diameter 26.2 mm; thickness 4.3 mm; 51 jewels; 28,800 A/h; 42-hour power reserve.

Functions: Hours, minutes, small second, chronograph.

Case: Matte stainless steel, diameter 44 mm, thickness 14.2 mm; sapphire glass; steel case back, six screws, transparent viewing aperture; waterproof to 5 bar.

Bracelet: Reptile leather with folding clasp.

Girard-Perregaux Monte-Carlo 1973

Movement: Self-winding; Caliber GP 337A; diameter 29.3 mm; thickness 8.5 mm; 63 jewels; 28,800 A/h; 46-hour power reserve.

Functions: Hours, minutes, small seconds, date, fly-back chronograph.

Case: Matte/polished stainless steel, diameter 40 mm, thickness 12.4 mm; sapphire glass; steel case back, six screws; waterproof to 3 bar.

Bracelet: Reptile leather with folding clasp.

Remarks: Limited to 250 watches.

TACHOMETER
MINUTES
SPLIT SECOND
CHRONOMETER
DATE
CHOPARD
hanhart
1882
SWISS
24
MADE

A Link to Motor Sports

Few other watch brands have devoted so much attention to classic motor sports as Chopard. In recent times Hanhart has also become seriously involved in this area. Not without reason, as classic automobiles provide beautiful backgrounds for beautiful watches. We have chosen two models that play with the classic motor sports colors red and black.

The Mille Miglia had a fixed place in the events calendar at Chopard. Since 1988, the Geneva watch and jewelry maker has sponsored the vintage long-distance race in Italy, where one can confidently speak of tradition. Because the passion for historic automobiles felt by Chopard bosses Karl and Karl-Friedrich Scheufele is so great, the company also supports the Grand Prix Historique, held in Monaco every two years. Finally, over several decades father and son have amassed a veritable collection of classic limousines and sports cars. With a consistency that is unique to this company and its leaders, they also dedicate their passion to watch lines like Mille Miglia, Grand Prix de Monaco Historique, and finally, Classic Racing Superfast, one of which is the Chrono Split Second, which we will examine in greater detail.

With it at the starting line is the Hanhart Primus Racer, which, like the Chopard chronograph, is finished in motor sports black with a red pushbutton (more on this later). With this watch Hanhart established a link to classic motor sports, where for many years the mechanical stopwatches from Gütenbach have been very popular. The traditional brand from the Black Forest, which celebrated its 130th birthday in 2012, maintains its tradition of time measurement at auto races by sponsoring various old-timer rallies, and since last year, has been present at a number of vintage auto shows in Germany. In 2011

The Chopard's split-seconds function is controlled by the red pushbutton. This enables the used to measure interval times, for example.

the watch brand supported the Rallye Solitude Historic in Stuttgart and the Jochpass Memorial in Bad Hindelang in the Allgäu District as chief sponsor and also sponsored important races like the Wurttemberg Classic in Garmisch-Partenkirchen and the AvD Oldtimer Grand Prix at the Nürburgring. Finally, in early 2012, Hanhart announced its association with the Cologne Porsche specialists Kremer Racing.

Classic motor sports or not—neither manufacturer is riding the retro-wave; instead, they are making highly desirable modern watches with which the wearer cuts a good figure, and not just behind the wheel of a vintage automobile. Both are black, sometimes shimmering in anthracite, and the stainless steel cases are DLC coated. The abbreviation stands for "Diamond-Like Carbon," which is described as being extremely scratchproof and durable. After weeks-long wearing tests we can confirm this. Another feature shared by both watches is a red chronograph button.

Red is the symbol of speed. All of the hands associated with stop-time measurement are painted red.

The La Joux Perret movement is hidden behind a massive steel case back.

While at first glance it looks like a pin clasp, it is in fact a folding clasp.

On the Chopard Classic Racing Superfast Chrono Split Second, this is almost exactly at 8 o'clock on the case and controls the split-seconds function. If it is pressed during micro-chronometering, a white seconds hand stops while the rest of the chronograph hands keep moving. During races, for example, this makes it possible to calmly measure and record lap times. Another press of the red button and the white seconds hand snaps back to its original place under the red chronograph hand.

The other two buttons for controlling the chronograph functions are formed more discretely and are not immediately recognizable as such, at least not optically. Like the crown, however, they are covered with rubber. Here the makers permitted themselves a little gimmickry: the crown adorns a star, which on closer examination resembles a perforated three-spoke steering wheel. Other automobile allusions are the "cooling gills" on the flanks of the 45-millimeter case and on the dial, plus the Allen bolts that fix both the bezel and the case back to the case center-section.

Chopard Classic Racing Superfast Chrono Split Second

Movement: Self-winding; La-Joux-Perret Caliber 8721; diameter 30.4 mm; thickness 8.4 mm; 27 jewels; 28,800 A/h; 46-hour power reserve; chronometer-certified (COSC).

Functions: Hours, minutes, small seconds, date, split-second chronograph.

Case: DLC-blackened stainless steel, diameter 45 mm, thickness 15.4 mm; sapphire glass; screw-down crown; waterproof to 10 bar.

Bracelet: Leather with folding clasp.

The bezel is provided with an engraved tachometer scale, which—because it is not color enhanced—displays very discreet harmony. If the observer's gaze moves towards the center, he discovers a Réhaut, whose scale is dedicated to indicating stop time—with raised numbers from 05 to 60, plus white dashes marking the seconds. The space between each set of second markings is filled by three red dashes, which, technically, is tailored to the oscillation frequency of the balance wheel in the Caliber 8721 from La Joux-Perret: 4 Hertz is equivalent to quarters or eighths of a second and not fifths or tenths of seconds.

In addition to the slave pointer chronograph function, this watch also has a date hand whose scale is slightly raised above the chronograph subdial. Finally, the movement, which is hidden beneath a steel case back and is based on a good Valjoux caliber, was also tested according to chronometer norm (COSC).

The Chopard's solid equipment includes a black leather strap, which in shape and volume fills the space between the two lugs and comes with a slight radius for better fit. The folding clasp ensures a firm fit on the wrist, and two red ornamental seams mirror the color of the chronograph hand and the third pushbutton.

The Hanhart Primus Racer also has a red pushbutton, however it does not control a slave hand, instead simply controlling the start-stop function. Red is not just loved by drivers of Italian sports cars, it is also the color of love.

According to a legend that Hanhart promoted in the 1930s, while putting on his Hanhart aviator's watch one morning, a young pilot discovered that his wife had painted one of the pushbuttons with red nail polish, so that he would always think of her and return home safely. Even if it isn't true, it is a good tale. Mind you the start-stop button on this Hanhart is anodized, not painted.

In general, the Hanhart is more reserved than the Chopard. The color red is only found on the tip of the stop seconds hand, n the corresponding null marking on the glass edge and as markers on the stop minute and the small seconds. Their embossed chrome outlines and the combining elements of the two subdials, which are also chromed, give the Primus Racer some cockpit appeal.

Otherwise Hanhart's artistic references to automobiles are rather sparse. Worth mentioning, perhaps, would be the Allen bolts that hold the flexible strap lugs in place. The latter are worthy of mention not only because of their artistic form, but also because of their ergonomics, ensuring that the partly-padded cowhide strap adapts to any wrist, no matter how large, after a brief wearing period. The effective strap length can effortlessly be adjusted at the folding clasp.

The Primus Racer is powered by a movement that Hanhart gives its own caliber number, HAN38. In fact it is an ETA 7750 (Valjoux) individualized with a special rotor. The movement can be seen through an aperture in the case back. This does not bother the watch as a whole, for it is very harmoniously formed. Optically, the conically-shaped center-section takes the weight that comes with the 44-millimeter diameter and 15-millimeter thickness. The serrated fixed glass edge comes from the company's own bag of tricks and is reminiscent of the knurled rotating bezels of old Hanhart aviation watches. It is all finished off by a pleasantly tidy dial. Our only wish would be for a slightly larger date window. And, as we mentioned the seconds scale on the Chopard's Réhaut, we do not wish to neglect to say that in the division into fractions of a second, Hanhart opted for user familiarity rather than the balance wheel's oscillation frequency (also 4 Hertz). Four intermediate dashes enable the display of fifths and tenths of a second.

But we don't want to be more Catholic than the Pope, and in the end both chronographs are really beautiful, well-made watches, which look unobtrusively masculine and sporty. Each customer must decide for himself if this is worth $5,500 for the Hanhart or almost three times as much for the Chopard. One thing for certain is that people with an inclination for sporty driving will enjoy either model.

Hanhart Primus Racer

Movement: Self-winding; Caliber HAN38 (ETA 7750); diameter 30.4 mm; thickness 7.9 mm; 28 jewels; 28,800 A/h; 42-hour power reserve.

Functions: Hours, minutes, small seconds, date, chronograph.

Case: DLC-blackened stainless steel, diameter 44 mm, thickness 15 mm; sapphire glass; steel case back with viewing aperture; screw-down crown; waterproof to 10 bar.

Bracelet: Calf's leather with folding clasp.

With just two subdials, the Hanhart's dial is characterized by calm and balance.

The high-quality folding clasp fits well into the overall image of the watch.

Hanhart individualizes the Caliber ETA 7750 with its own rotor.

Pilot's Watches

OFF
HYD
BAT
FAULT RESET
TUCSON
NEW YORK
CHICAGO
FAIRBANKS
LOS ANGELES
CARACAS
SAO PAULO
AZORES
WEST
LONDON
BERLIN
HELSINKI
MOSCOW
DUBAI
KARACHI
DACCA
BANGKOK
HONGKONG
TOKYO
SYDNEY
NOUMEA
AUCKLAND
HONOLULU
MIDWAY
GLOBAL-TIMER

High-flying Wrist Instruments

What is a true pilot's watch anyway? Modern pilots discuss this just as frequently and passionately as watchmakers. The least common denominator is usually this: a pilot's watch must be a highly accurate timepiece that is easily readable even under adverse conditions. Such watches also have a whole series of features that are useful both in an aircraft and everyday use.

The jewelry- and watchmaker Cartier is not actually considered a specialist in sports watches. And yet the first timepiece to earn the title pilot's watch came from that very house. During a dinner in Maxim's, Alberto Santos-Dumont, known in the Paris of the Belle Epoque as a dandy and aviation pioneer, complained about the problems he was having with his pocket watch. Because he literally had his hands full during experiments with his hydrogen balloons, he simply had no chance to check flight times. At that time pocket watches were worn on gold chains. Pulling it out of his pocket, reading it and putting it back again could result in a fatal loss of control. Louis-Joseph Cartier, a friend of Santos, listened attentively—and with master watchmaker Edmond Jaeger developed the first wristwatch. It had Roman numerals, two hands, and a handy crown and was attached to the wearer's arm by a leather strap.

Stowa makes a modern interpretation of its classic pilot's watch. The new version comes in a 40-mm stainless steel case and, if desired, with a pilot's strap.

This look back reveals characteristics that are still of significance today. The first of these is a firm attachment to the wrist, such that the watch can always be seen and read. In some cases, pilots in the First and also the Second World War had their watches fitted with overly-long leather straps which enabled them to be worn over heavy flying jackets or overalls—and if needed even over the thigh.

The second characteristic is quick readability. Even in the digital age, an analogue watch has clear advantages over a digital one. We can interpret the position of two hands in fractions of a second, while a digital indication literally has to be read. The military sees it the same way. Good readability is priority number one. For this reason, most watches characterized as pilot's watches have black dials with white numbers and hands, which, for better night readability, have luminous coatings and a specially marked 12 o'clock position. A prime example in this respect are the so-called "observation watches" used by Luftwaffe

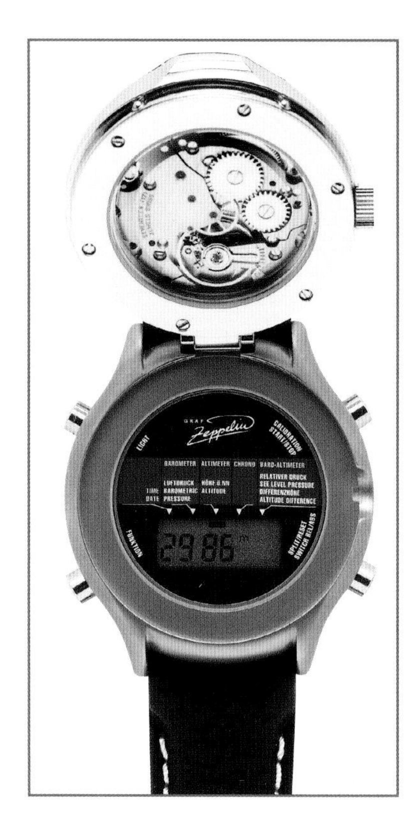

Above: PointTec in Munich specializes in moderately priced pilot's watches. Here is a Zeppelin with manually wound mechanical movement and electronic altimeter.

Center: The Breitling Emergency is equipped with a micro-transmitter that transmits on the aviation emergency frequency 121.5 MHz.

Right: The IWC Mark XI was used as the official service watch by the Royal Air Force, but was also used by civil pilots.

pilots, also called "B-Watches" for short. They were made in the 1940s by Wempe, A. Lange & Söhne, IWC, and the Pforzheim companies Stowa and Laco. The last three have modern interpretations of their former "professional watches" in their collections.

Strictly speaking, since the Second World War, magnetic field protection for the movement has been an indispensable feature of a watch. The screens of increasingly common ground-based radars produce magnetic fields of such strength that the running of a watch can be significantly affected. Alloys less sensitive to magnetism are used in the area of the works, especially on the hairsprings, yet even today, magnetic fields in cockpits can negatively affect a watch. Makers of pilot's watches who follow the pure doctrine—like Sinn and IWC—still equip their timepieces with inner cages of soft iron, which direct magnetic fields around the movement.

For pilots and "seconds foxes" among watch lovers, stop-seconds are an indispensable feature. This is understood to mean a mechanism that stops the watch when the winding crown is pulled and starts again when it is pushed in. Watches can thus easily be synchronized, which is indispensable, especially during military operations.

While it is generally dry in cockpits—prewar machines excepted—a pilot's watch frequently comes in contact with water before and after a flight;

consequently, water resistance is an indispensable quality as well. But beware: the information on the watch should not be taken literally. A watch that is DIN rated watertight to 30 meters is immune to splashed water, and hand washing or a walk in the rain are no problem. But anyone who wants to wear a watch while swimming should select one that is watertight to at least 100 meters (DIN standard).

The quality of the watch glass plays a role in the everyday life of a pilot. The Omega Speedmaster Professional, dubbed the Moon Watch because of its use on several Apollo missions, was and is equipped with an acrylic glass lens. Thanks to its elasticity, it can adjust to pressure differences without popping out or shattering. It is scratch-sensitive, but unlike mineral glasses, scratches can be polished out. The first choice in high-quality watches today is sapphire glass, which is anti-reflection coated on one side. These watch crystals are scratchproof and offer good readability.

Among the features most prized by pilots today is the stop function in chronographs for micro-chronometering up to twelve hours. The more advanced choose trailing pointer chronographs with two stop-seconds hands. One can be set at an interim or reference time while the other continues to run. This is what is meant by the term "split seconds." If the reference time is no longer needed, a simple push on the button is all that is needed to cause the second hand to jump to the position of the first. It catches up, which is why the function is also called *rattrapante* (*rattraper* means "to catch up to" in French). The flyback function is important to many pilots. This "reset in flight" makes it possible to zero the chronograph and immediately restart it without activating the start-stop button. One can ring in a new measuring interval

Because of its use by NASA, the Omega Speedmaster Professional is also called the "Moon Watch." In addition to the classic, Omega regularly offers special models.

with the touch of a button. Active pilots, especially those who move between different time zones, are predestined for watches that can display at least two times at once (for example, home and local time, or Zulu and local time).

While the wristwatch was an indispensible instrument in the early days of aviation, today it usually has a very different function. A Lufthansa captain philosophized: "The more digital and highly technical our environment becomes, the greater the desire for something tangible and down to earth, which a mechanical watch embodies. It represents an endearing contrast to our fully electronic working environment. For me that is its special charm."

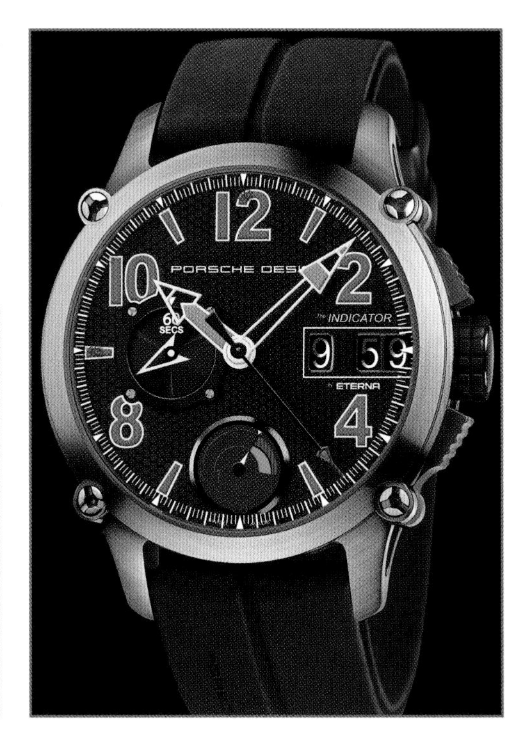

Above left: Pilot's watches have soft iron cages for protection against magnetic fields. The iron cover under the steel case back is also part of the anti-magnetic shielding.

Above right: The Porsche Design Indicator has a digital chronograph display.

Above center and bottom: The TeStaF testing program (see box right) places watches and their components under chemical and mechanical stress.

The Sinn Pilot's Chronograph 756 UTC prior to assembly.

Technical Standard for Pilot's Watches

Takeoff clearance issued

At the initiative of Sinn Special Watches, a project group from the University of Applied Sciences in Aachen defined criteria that a wristwatch must meet to be considered a pilot's watch. This Technical Standard for Pilot's Watches was issued takeoff clearance at Eurocopter in Donauwürth in the summer of 2012.

"What must a watch be capable of doing, in order to bring me safely home after the loss of my instruments?" This question occupied Professor Dr.-Ing. Frank Janser, head of the Air and Space Technology Department at the University of Applied Sciences in Aachen—and a passionate pilot. He formed a project group of specialists from his institute and the watchmaker Sinn Special Watches of Frankfurt. It had long been a thorn in the side of the company's owner, Dipl.-Ing. Lothar Schmidt, that while there were two standards for diver's watches (DIN 8306/ISO 6425), there was none for pilot's watches. In Janser's institute he found an ideal partner, as the air and space technicians were not only intimately familiar with aircraft but with their instrumentation or avionics as well.

The project group questioned active pilots and examined aviation and watch standards, and from this developed a catalog of criteria that defined what made a usable pilot's watch for active pilots. The three key points were functionality (for example, readability by day and night, accuracy, and power reserve), resilience (for example, pressure resistance, operative temperature range, G-loads, water resistance, shock resistance), plus safety and compatibility (no effect on avionics, secure bracelet attachment). The detailed description may be seen at www.testaf.org.

During a type-examination, for which the FH calculates a cost of about $11,000, the entire catalog of criteria is gone through. The manufacturer has to submit two complete watches, a complete watch case, nine watch crystals, and nine seals. If successful, the watch is certified, and watches of that type can then bear the TeStaF seal and be delivered with a corresponding certificate. The first two watches to pass the TeStaF test are the Sinn Model 103 and the EZM 10. Others will surely follow.

BREITLING
1884
NAUT.
STAT.
MPH

Cockpits on the Wrist

People on trips, especially pilots, value watches that show both the local time and the time at home. The Breitling brand, with its close association with aviation, offers several, from the classic Navitimer World to the racy Chronomat GMT to the technically clever Transocean Chronograph Unitime.

Just prior to landing, say after a trans-Atlantic flight, the fumbling begins: hectic turning of the crown to set the new local time on one's watch. But then, what time is it at home? Is it too late to call? The owner of a GMT watch can answer such questions easily and with a smile, after all he is master of two or more zone times.

Yes, they are called *zone times* and not *time zones*. The latter term describes a geographical area that has a common time of day and a common date. The story behind it: in the mid-1900s, the growth of ship and rail traffic required a coordination of times, which often differed by several minutes from place to place. After protracted international negotiations, in 1884, an agreement was reached at the Meridian Conference in Washington, D.C., to divide the earth into 24 times zones, each with 15 degrees of longitude. The zero meridian was defined as the degree of longitude passing through the English city of Greenwich. Instead of GMT, or Greenwich Mean Time, which was used until the 1970s, today we use UTC, or Universal Time Coordinated, which is also the international standard for aviation. The time in Germany is therefore UTC + 1 (plus one hour), and UTC + 2 in the summer. Nowadays local time is not just determined by degree of longitude, but by political considerations as well. China, which, because of its vast east-west domain, should have five time zones, actually has just one, the local time in Beijing. And many nations, such as India (UTC + 5.30 hours), even use half-hour increments. The majority of nations, however, limit themselves to increments of an hour.

The Breitling Chronomat GMT is equipped with the manufacturer Caliber B04.

Breitling chronographs are helpful in this respect, as they can all indicate a second zone time while having significant differences functionally. Let us begin with the classic, which celebrated its 60th birthday in 2012. It is the Navitimer, the watch with a slide rule incorporated into its rotating bezel. With its logarithmic scales, the slide rule is meant to assist in

For advertising purposes Breitling maintains an entire fleet of aircraft, including biplanes, aerobatic machines, a vintage Super Constellation passenger aircraft and several jet trainers.

navigation and, of course, serve as a timer, hence the name Navitimer. The Navitimer World (starting price about $7,100) has a red-tipped 24-hour hand. It is powered by a Caliber 24, which is nothing more than a refined ETA-Caliber 7754. A feature of this watch is that only the 24-hour hand can be set to hourly increments, while the watch keeps running. As on other watches, the local time is set using the minute and hour hands. This is somewhat cumbersome on long trips.

Somewhat better in this respect is the Chronomat GMT (starting price about $9,700), which also has a waterproof in-house movement. But let's start at the beginning: in the summer of 2004, under strict secrecy, development began of a new chronograph movement that was supposed to be "100% Breitling." As noted earlier, equally important to the watchmaker that sells more than half a million timepieces per year, the movement had to be produced industrially on a large scale at an acceptable cost and consistent high quality. Precision is ultimately a part of the self-image of a brand that sends every watch movement to the COSC for chronometer testing. With the help of outside specialists, a development process proceeded swiftly: at BaselWorld 2009, on the brand's 125th birthday, Breitling unveiled the first chronographs with its own Caliber B01—a three-point landing.

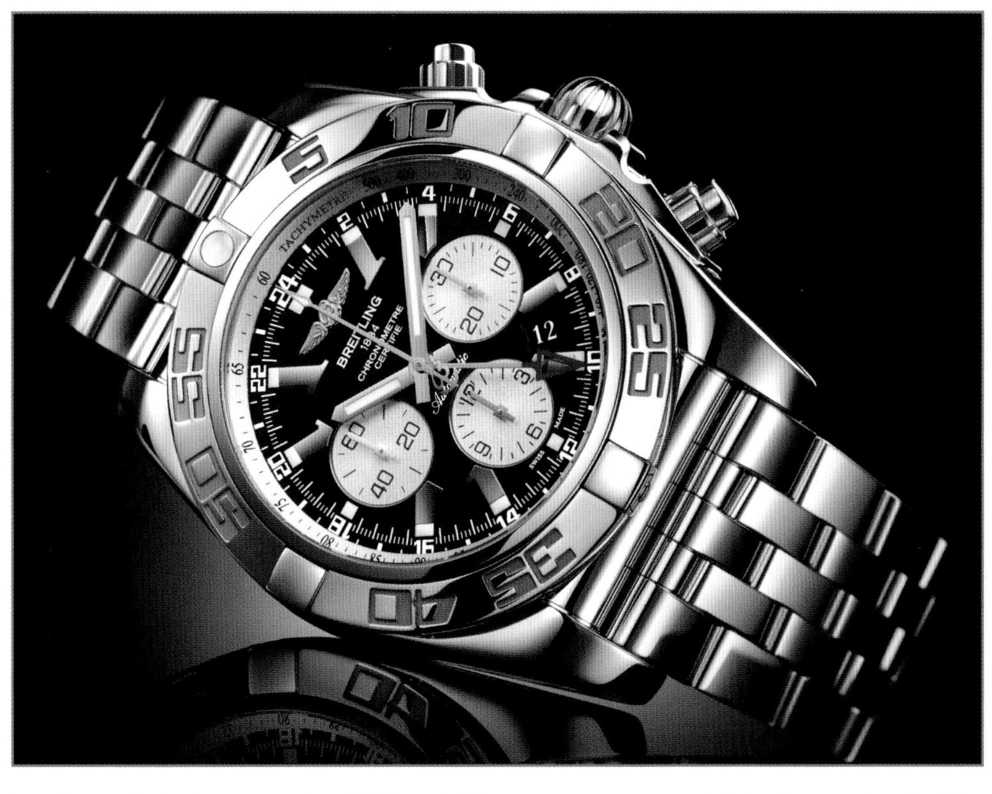

The Breitling Chronomat looks like a real fireball, but it sits very comfortably on the wrist.

For fans of technology: the Caliber B01 is a column-wheel chronograph with automatic winding and a vertical clutch between the motor and stop mechanism. The stopwatch's hour counter turns continuously, whereas on many classic chronographs it skips. Other special Breitling features are the patented auto-centering mechanism and regulation of the watch by means of a rotating body in the balance cock.

A GMT function has since been added to this movement. Essentially it consists of a 24-hour hand plus a separately adjustable hour hand. The red-tipped 24-hour hand marks home time, or, in the case of UTC pilots, the small hour hand shows the local time. The latter can be adjusted by a turn of the screw-down crown. The hour hand then jumps forwards or backwards as required. If, in the process, it crosses the date limit, the date indicator also resets itself. This is valued not only by pilots, among whom Breitling is highly esteemed, but by passengers as well. The former also find the rotating bezel with its sixty graduations a useful tool, while for the normal wearer the prominent stainless steel ring is simply an adornment that underlines the instrument character of the watch.

Despite all its technical appeal, the Chronomat is a stylish watch whose face is dominated by the attractive dial. It appears in fine anthracite, the three subdials in plain black. The hands and markers, which are polished to a high gloss, contrast well with the dial and provide a high level of readability—even at night thanks to luminous markings. The Breitling can definitely be worn with a suit, although, with a diameter of about 46 millimeters and a height of roughly 18 millimeters, it is not exactly cuff friendly. With its rubber strap and finely adjustable clasp, the watch is so comfortable that the wearer will rarely want to take it off, which, thanks to its ability to withstand 50 bar of pressure (500 meters of water), is almost unnecessary.

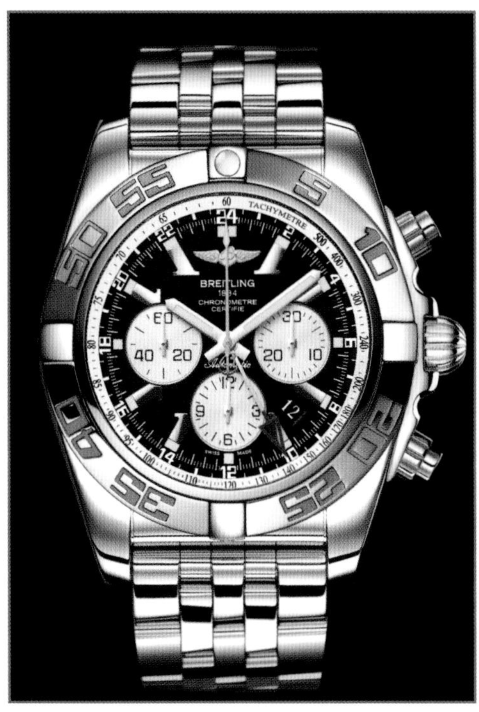

Breitling stresses its closeness to the aviation world by operating a squadron of aircraft that regularly appear at air displays. Here are two biplanes with wing walkers.

The Chronomat GMT is available in a number of versions. Above right is the limited edition with 24-hour bezel.

Just in time for BaselWorld 2012, Breitling was waiting with a watch for frequent flyers. The Transocean Unitime model is a world time watch—simple and logical to operate. With a turn of the crown, the zone times can be set in the blink of an eye. The time shown by the two central hands is always that for the city at the top in the 12 o'clock position; all others can be read using the 24-hour ring. During a time zone change, therefore, all that is required is to pull the crown and set the time forwards or backwards in one-hour increments. The date changes when the date limit is passed and always matches the time zone shown at 12 o'clock. This patented mechanism is part of the new in-house B05 movement, which may be seen as a development of the B04 from the Chronomat GMT. The stainless steel version of the Transocean can be had for just under $14,000, while the red-gold version costs more than $33,000.

The Transocean Unitime is aimed more at air travelers than pilots. The world time watch is simple to use, but the dial looks too busy for an instrument.

IWC
IWC
SCHAFFHAUSEN

Legendary Pilot's Watches

Those who wish to manufacture pilot's watches must know something about flying. Since the 1930s, IWC Schaffhausen has made professional timepieces adapted to the special conditions of aviation—for pilots and passengers. Maintaining this tradition, IWC continues to make authentic, modern pilot's watches that are reliable companions for world travelers.

Flying symbolizes modernity and freedom. Many aviation pioneers had to risk their lives to reach this point, however. Pocket watches attached to an instrument panel provided less than optimal timekeeping in the early days of aviation. The conditions of a cockpit—vibration, extreme temperature variations, and powerful magnetic fields—required a special watch for airmen. And—within limits—that is exactly what the Mark IX of 1936 was, with its black dial for optimal readability, prominent illuminated hands and large illuminated numbers, a rotating bezel, and a shockproof, antimagnetic escapement.

It was followed in 1940 by the Big Pilot's Watch designed to military requirements with original pocket watch movement and large second hand—a certified observation and navigation watch for combat pilots. The most famous IWC pilot's watch, the

Aviation pioneer and author Antoine de Saint-Exupéry endorsed watches for IWC Schaffhausen. Several special models were dedicated to his memory.

The Big Pilot's Watch (left) is very popular, and not just among fans of IWC watches. The Hand-Wound Pilot's Watch (right) is part of the Vintage Collection unveiled in 2008.

Mark XI with the manually wound Caliber 89 movement, entered civilian and military use in 1948. For a long time, the Mark XI was the official service watch used by pilots of the Royal Air Force, but was also worn by civilian pilots of BOAC (British Overseas Airways Corporation) and other airlines. Its advantage compared to other pilot's watches: it had an additional inner case of soft iron for magnetic field shielding and achieved such a high level of perfection that, 55 years later, its escapement was adopted piece-for-piece in the second Big Pilot's Watch with the seven-day automatic movement. The Big Pilot's Watch of 1940 and the Mark XI of 1948 are still running today, but most have found their way into the jewelry boxes and safes as rare, high-end collector's items.

As flying became more matter-of-fact, pilot's watches went out of style for a time, at least among passengers. That is until 1988, when IWC rediscovered its glorious tradition as a maker of pilot's watches. They debuted a pilot's chronograph with quartz movement. Its successor was equipped with a mechanical movement, and that was followed by the IWC Double Chronograph with automatic movement and rattrapante function, which in 2006 took off with a high-tech ceramic case.

In 1994, those who missed out on the legendary Mark XI welcomed the arrival of the more up-to-date Mark XII—now with automatic movement and date indicator. In keeping with the public's tastes, it was later made larger as the Mark XV and Mark XVII and today forms the basis of the

In 2004, IWC enhanced the classic pilot's watch look with the Spitfire models' silver three-dimensional-looking dials.

For pilots and collectors: the Double Chronograph in ceramic case (left) and the Pilot's Chronograph Edition Saint-Exupéry.

An engraving adorns the Saint-Exupéry special limited edition.

The Spitfire pilot's watch and the British fighter aircraft that gave it its name. The aircraft's former military role was of secondary importance to IWC. Instead the Swiss emphasize the undoubted elegance of the aircraft.

The Pilot's chronograph comes in many versions: the limited Edition Saint-Exupéry in three house colors (left) and with Spitfire dial (right).

IWC pilot's watch family. The Pilot's Watch Chronograph Automatic, the Pilot's Watch Double Chronograph, and the UTC all share the ingenious feature of adjusting time zones forwards or backwards one hour at a time by turning the crown, which for pilots solves the confusion of 24 time zones in the simplest way. It is also possible to select home time, a second preferred time or simply UTC, the reference time in aviation the world over, displayed in a small window.

Since 2004, other pilot's watches with a different look and named after elegant aircrafts, have also been available. The series is called Spitfire. The series looks less robust than the classic pilot's watches with their resemblance to instruments. According to IWC, the goal is to reflect its namesake aircraft's outstanding technology and cool elegance rather than a military purpose. Just as panels on aircraft are usually attached to the framework with rivets, on the new Spitfire watches the numbers and indexes are riveted to the dial. They are grouped around a sublime, distinctive dial center. The surface areas of the dials, which are given a semi-matte finish in an electroplating bath, contribute to the elegance of the Spitfire collection.

Since 2002, the flagship has been the Big Pilot's Watch, an unmistakable timepiece with a diameter of 46 mm, thickness of 16 mm and, in its steel version, a gross weight of 150 grams. Everything that is good and expensive, and which has proved itself magnificently in IWC's long history of making mechanical watches, comes together in the in-house Caliber 5011. At present, it is the largest automatic movement in the world. Using the patented pawl winding system (Pellaton system), it quickly produces the energy for an 8.5-day power reserve; however, it provides exactly 168 hours, or seven days, before the power reserve's sophisticated drive mechanically stops the movement. In this way, diminishing torque in the mainspring, which can cause a loss of amplitude in the balance wheel and thus errors, is eliminated.

IWC's patented pawl winding system was developed by former IWC head Albert Pellaton in the late 1940s for the legendary Caliber 852 family and their successors 853, 8531, 854, 8541 and 8541 BS. Used successfully for decades in a wide variety of IWC watches, such as the Yacht Club, Engineer, and Golf Club, at the end of the millennium, it experienced a much-deserved renaissance with the watchmaker's Big Caliber 5000, and the Caliber 5011 now powers the Big Pilot's Watch.

The Big Pilot's Watch is also very well equipped externally for all aviation and also civilian tasks: its indications readable in every situation and in any light behind the convex, anti-glare sapphire glass. Features include the central seconds hand, a naturally large date indicator, and an indicator for the movement's seven-day power reserve. Like all mechanical pilot's watches from IWC, the Big Pilot's Watch has the special features associated with the highest level of flight capability. These include the soft-iron inner case, the special glass seat against a sudden loss of pressure in the cockpit, the screw-down crown, and the tested water resistance to 60 meters. The Big Pilot's Watch Perpetual Calendar represents the current apex of the series. It is an imposing timepiece with a 48-mm case made of ceramic and titanium, equipped with the patented perpetual calendar of the former head of development Kurt Klaus.

The Double Chronograph with ceramic case and the current dial—recognizable by its larger date window.

Sinn
Sinn

Sinn's Half-century Mark

Sinn survived a wave of mergers in its sector of the watch industry just as it had survived various other crises. Dependable quality and a good price-performance ratio replaced fat marketing budgets. Since its founding in 1961, the owners of Sinn Special Watches of Frankfurt am Main have driven the agenda—first pilot Helmut Sinn, and since 1994, engineer Lothar Schmidt. With unconventional ideas, plans, and projects, the medium-sized watchmaker has now been in the business for more than half a century.

A mechanical watch movement is a highly complex and at the same time highly precise machine. It achieves a degree of accuracy of 99.9 percent if it is ten seconds fast or slow per day, which, by the way, is the absolute limit for Sinn watches according to internal quality standards.

It is therefore not surprising that the watch movement must serve as the symbol of precision and reliability. In common parlance, companies that continually bring good products to the market are said to "run like clockwork." The watchmaker Sinn, based in the Frankfurt district of Rödelheim, founded in 1961, is surely one such company. Pilot and flying instructor Helmut Sinn began producing watches there under his own name because the good pilot's watches on the market were too expensive and the cheaper ones seemed to be of poor quality. He had his watches made by Swiss watchmaker Guinand according to his wishes and specifications and inscribed them with his company name—creating a private label, which is not necessarily a bad thing.

As a pilot, Helmut Sinn was aware of the importance of onboard navigation clocks, which for a long time were a significant source of income for the company. These clocks were installed in the cockpits of many military aircraft, but also in the Boeing 727 passenger jet. Until 1980 the ratio of onboard navigation clocks to wristwatches was about 4:1. At the beginning of the year, however, Sinn Special Watches encountered a supply shortage: the mechanical movements for the navigation clocks were no longer obtainable. The company

The Founder

Helmut Sinn was born in Metz in Alsace-Lorraine in 1916. He grew up in the Pfalz, became interested in aircraft, and subsequently attended a flying school. His graduation coincided almost exactly with the outbreak of the Second World War. In the years that followed, he became an instrument and aerobatic flying instructor and got to know and value of precise, reliable aircraft instruments. After the war, Sinn first established a watch wholesaler and was a rally driver in his spare time. In 1961, he founded the watch brand Sinn, which he sold to Lothar Schmidt in 1994. His planned retirement was short-lived, however. In 1995 he took over the Swiss watchmaking company Guinand, which had made Sinn watches as private.label products until 1994.
In 1996, at the age of eighty, he founded a new company, and with mild self-irony, named it Jubilar (birthday boy). Its line included Jubilar pocket watches and Chronosport wristwatches, and later Guinand wristwatches as well. Today, the Frankfurt company operates under the name Guinand Watches Helmut Sinn GmbH (Ltd.), and the managing director is Horst Hassler.

The mechanical navigator's clock—NABO for short—was mounted in the cockpits of commercial aircraft and can still be found in general aviation aircraft.

Scientist-astronaut Dr. Reiner Furrer went into space with his privately purchased Sinn 140 S and worked with it in the Spacelab.

In 2005 Sinn celebrated the anniversary of the Spacelab mission with a special series restricted to 500 watches.

needed an alternative and invested in the development of an onboard clock with a quartz movement—unfortunately without success. The prototypes failed tests by the Federal Office for Defense Technology and Procurement because they could not withstand temperature shocks between +60°C (140°F) and -35°C (-31°F). There were also problems during the dust and salt spray tests. Sinn Special Watches therefore pushed deeper into the field of wristwatches.

In the process, it received unexpected and free publicity. In 1985, the German physicist and astronaut Dr. Reinhard Furrer purchased from Sinn the Model 140 S with the Caliber Lemania 510. No one in Frankfurt had any idea what was going to happen to the watch. Furrer wore it on his wrist during the D-1 Spacelab mission—which meant a real world premiere. It was the first time that a chronograph with automatic winding encountered weightlessness in the vastness of space. That was unusual, but until then, watch experts had assumed that automatic wristwatches would not work, as the absence of gravity would prevent the rotor from carrying out its task of providing energy. Reiner Furrer showed that the contrary was true. The watch worked during the trip at an altitude of 383 kilometers (238 miles), covered 4.6 million kilometers (2.9 million miles), and performed flawlessly. Sinn loved this kind of free advertising—and still does.

Streamlined cost structures were a specialty of Helmut Sinn. There was no real advertising budget and no dealer margins. At the time, the company's products were sold exclusively from the factory. The only advertising was word-of-mouth among career and private pilots. News of the brand's high functionality and quality quickly got around. An important argument for Sinn watches has been their good price-performance ratio, for the Frankfurt company has been one of the few to sell watches through direct sales.

That remains true to this day, though a few things have changed at Sinn. In 1994, the company's founder and name-giver sold it and the brand to Lothar Schmidt. Since then, Schmidt has been Sinn's mainspring in its efforts to remain a player in the watch business. With a degree in mechanical engineering, he drives the agenda in Rödelsheim—and successfully. In a very short time, he doubled the company's sales and earnings, in part because he gained the support of the specialized trade. In 1995, the jeweler Oeding-Erdel opened the first so-called Sinn Depot, where customers could purchase Sinn watches at factory prices. The system worked. From 1994 to 2010, Sinn increased its staff in Frankfurt from 14 to more than 70, and today, Sinn Special Watches sells about 12,500 nwristwatches per year.

The Owner

Lothar Schmidt was born in Neunkirchen in the Saarland in 1949, completed an apprenticeship as a toolmaker before becoming a mechanical engineer and, in a second course of studies, earned the title of REFA-Engineer. After serving for two years in the military, he began his professional career as a design engineer with a Swiss engineering factory and later transferred to the watchmaking sector. Lothar Schmidt began his activities with IWC Schaffhausen in February 1981, first as a freelancer, and later as an authorized signatory. He was initially responsible for the establishment and operation of watch case and bracelet production and for their design and development. He later also took over the establishment and operation of movement parts production. From 1990 until he left IWC in 1993, he also headed the production and logistical development of the reestablished A. Lange and Söhne GmbH, then a subsidiary of IWC. During the takeover preparations, he was already employed by Helmut Sinn GmbH (Ltd.) from September 1993 to August 1994. Since then he has headed Sinn Special Watches as managing partner.

But don't believe that Schmidt spent a great deal of money on full-page advertisements and other advertising measures. That is not in keeping with his nature. The highest levels reached by the businessman were small format ads in special-interest media and, with an eye on the flying customer, sponsorship of an aerobatic pilot. The following is true: "The products must speak for themselves, and then success will follow." Schmidt says that he prefers to invest the money saved on advertising on developing his products. This notion, of course, spread among the constantly growing community of watch-lovers and may thus be considered a very subtle form of advertising.

Schmidt shows that these are definitely not just empty words. With satisfying regularity he and his team develop new specialties and ready them for production like no other watchmaker. His two diving watches provide an example. There is the Hydro, a quartz watch, which is completely filled with highly-purified silicon oil. Because the center part of the case back is designed as a membrane, the watch can endure any ambient pressure and is capable of handling any depth that can be reached by a human diver. A welcome side effect: because of its oil filling, underwater the Hydro is readable from any angle, while air-filled watches reflect to varying degrees because of the changed refraction between water and air.

The Arktis diving chronograph profits from another sensible idea by Schmidt, which, like the technology of the Hydro, he has had patented. Because humidity trapped during assembly and dampness diffusing through the seals can negatively affect the oil quality, Sinn fills the Arktis with argon gas, whose large molecules keep the damp ambient air away. A desiccant capsule screwed into the case also helps. It is filled with copper sulfate, which absorbs moisture and shows this by its color: in a dry state it is almost white, but when saturated with moisture it is a brilliant blue. This is a signal to the owner that his watch needs to be taken in for service quickly. Schmidt is so confident in this technology that he has increased the warranty period for these watches by a year.

The native of the Saarland made a name for himself in the field even before he bought the Sinn company, during his time as head of production at the Swiss luxury watchmaker IWC in Schaffhausen. He began there in 1980 as a freelancer and, over the years, worked his way up to authorized officer. And instead of going on vacation, Schmidt studied to become an REFA engineer. This readied him to take responsibility for the conception, planning, and installation of the production technology on behalf of IWC during the rehabilitation of the watch brand A. Lange & Söhne in Glashütte.

Dipl.Ing. Lothar Schmidt is the current owner and managing director of Sinn Spezialuhren of Frankfurt.

During his time at IWC, Schmidt was valued for his know-how in building cases, particularly those that used problematic materials like titanium, platinum, and ceramics. A milestone from this period, a Da Vinci with a black case, resides in his safe. The case is made of zirconium oxide, which is considered one of the hardest watch case materials. This watch inspired Schmidt to think about harder and therefore more scratchproof steel cases at Sinn. The first result was the Duochronograph 756 pilot's watch, whose steel case has a value of 700 on the Vickers hardness scale. To clarify: that is three to four times the normal. The material resistance is achieved through a bath in liquid nitrogen (-196 degrees Celsius). But that was not enough for the small businessman.

Engineer Schmidt once again put himself a nose-length ahead of the competition with technology known as Tegiment. Through a special procedure, only the external surfaces of a metal are hardened. This protective coat (Latin: *tegimentum*) produces a hardness of about 1,000 Vickers. "Resistance to scratches as well as corrosion resistance are both further improved", says Lothar Schmidt. "The Tegiment case can also be a blessing to allergy sufferers. The transfer of nickel to the skin of the wearer is reduced to almost zero." He says this with a little pride in his voice, but on the other hand he does not forget his fellow campaigners: "During development we worked very closely with Dekra Materials Testing in Saarbrücken, among others." By the way, the ambitious project was implemented by the case maker Saxon Watch Technology Glashütte (SUG), which Lothar Schmidt founded in 2000 together with Walter Fricker of Pforzheim and Dr. Ronald Boldt of Saxony.

This clearly illustrates Saarlander's way of thinking. He has no desire to reinvent the wheel. Instead he sounds out those who have already encountered a similar problem in another area. Then he seeks a competent comrade-in-arms and tries to adapt the solution to the problem to watchmaking. What sounds so simple is ultimately hard work, and time consuming, as the Diapal project shows.

The Sinn 903 Klassik chronograph has a modern phase of the moon display and an internal rotating bezel.

Sinn also delved into the technology of the movement and considered improvements to a system that had worked for over a century: the Swiss escapement mechanism. It has one serious weakness: "It requires constant lubrication to reduce pallet face and anchor pallet friction. Because the oil is subjected to an aging process, it has to be regularly maintained. Old oil in the area of the escapement causes running irregularities, and if there is no lubrication sooner or later the watch stops," explains Lothar Schmidt. The only person to have so far found a solution to this problem was the British watchmaking genius George Daniels. He developed the co-axial escapement, which is used by Omega. The basic principle of the design is to avoid longer sliding distances between the pallet and escapement wheel.

The modern 900 Flieger (middle) is framed by the perennials 103 Klassik (left) and 103 A.

Sinn took a different path. In Frankfurt, they sought a combination of materials which, in laymen's terms, rub as little as possible. Schmidt described the beginning of the project: "We began our first experiments in 1995, replacing the usual synthetic ruby anchor pallets with diamond pallets." This also gave rise to the name Diapal Technology. It failed to produce the desired results, however. Schmidt therefore began looking for a comparable problem and found his answer in spaceflight technology. After much research, he determined that: "Oil also cannot be used as a lubricant in space, which is why research has been done into material pairings for moving parts that have the lowest possible coefficient of friction." Schmidt and his technicians used these research findings and discovered a combination of escapement wheel and anchor pallets, which now in fact run without lubrication.

Schmidt not only seeks frictionless movements on the small scale but also on a larger scale, with the entire Sinn

As a pilot, Helmut Sinn emphasized the instrument character of his watches. Even then, however, the Frankfurt watchmaker also produced elegant timepieces.

company. While he would like to take care of every detail himself, he cannot, and so he sought staff. Not just employees, but people to think and work with him and keep the Sinn movement running. The key positions are held by highly qualified academics and craftsmen: engineers, physicists, management experts, and, of course, master watchmakers. Because Schmidt knows all too well that a watch movement only functions as well as its supposedly weakest part, he selects new personnel carefully. And in this he has had an extremely lucky hand. Several times, Sinn apprentices have been judged the best in the country in watchmaking and engraving. There is no cause for concern about the future.

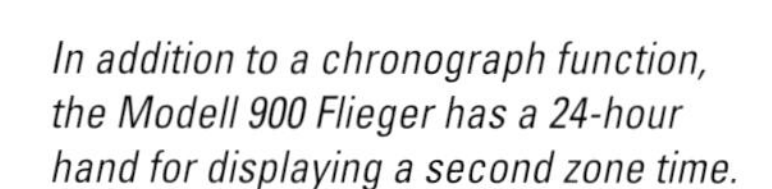

In addition to a chronograph function, the Modell 900 Flieger has a 24-hour hand for displaying a second zone time.

Sinn's Time Line

1961

In 1961 Helmut Sinn, an instrument flight instructor and pilot during the Second World War, established Helmut Sinn Special Watches in Frankfurt am Main. The company concentrated on the manufacture of on-board navigation clocks and pilot's chronographs—in large part because of the founder's biography. Though made in Switzerland, they were marketed as a private brand.

1985

Sinn Special Watches finds a new headquarters: from Rödelheimer Parkweg 6, it moved into new business quarters. The new address, where the company remains to this day, was just a few steps away: Im Füldchen 5 -7. In 1985, during Spacelab mission D-1, the German physicist and astronaut Professor Dr. Reinhard Furrer wore the 140 S with Caliber Lemania 5100 on his wrist.

1993

Cooperation developed between Bruno Belamich, who later became top designer at Bell & Ross, and Sinn. From that time onwards, selected Sinn watches were marketed by Bruno Belamich in France, the USA, and other countries under the name "Bell & Ross by Sinn." The business relationship continued until the end of 2001.

1994

At the age of 78, Helmut Sinn sold the company to engineer Lothar Schmidt. It marked the beginning of a new era for Sinn Special Watches. Schmidt wanted to give the brand more technical autonomy.

1995

The company unveils its first gold watch. Despite its unusually high gold content of 22 karats (917/1000) the metal's material hardness was equivalent to that of stainless steel (220 HV).

Sinn also presented dehumidifying technology using argon gas for the first time in the 203 Ti Ar. This improved its resistance to fogging and was also supposed to slow the aging of technical components. The warranty period for this watch was lengthened to three years.

Lothar Schmidt supplemented his direct sales with so-called depots, where watches were sold at direct-from-the-factory prices. The first official Sinn depot was opened by Oeding-Erdel jewelers in Münster.

1996

Unveiling of hydro-technology in the 403 Hydro EZM2 quartz watch, which is used by the maritime units of GSG 9, among others. With its oil-filled case and volume-compensating case back, Sinn guarantees that the watch is shockproof, pressure resistant at any depth that can be reached by divers, and 100% readable underwater from any viewing angle.

In the Watch of the Year selections, Sinn takes fourth place with the 203 Ti and seventh place with the 244.

1998

A practical test for the further-developed argon dehumidifying technology: the 303 Kristall is worn by five contestants in the Yukon Quest dogsled race in the freezing Yukon and Alaska at temperatures of -40° C (-40° F). The watches, worn under their warm clothing, perform flawlessly.

1999

Lothar Schmidt and Dr. Ronald Boldt form a new company, Saxon Watch Technology GmbH (Ltd.) Glashütte. The company produces watch cases for Sinn Special Watches that are technologically demanding to make.

The Frankfurt Financial Center Watch is unveiled. This watch is the first in a line of elegant sports watches that is developed in the years that follow and today includes eight different watches. The legend "Frankfurt am Main" appears on a Sinn watch dial for the first time.

2001

Sinn Special Watches turns 40 years old. The company produces an anniversary catalog to mark the occasion. For the first time, the catalog appears in book form with a rigid spine.

The Diapal Technology, in which ruby pallets are replaced by diamond ones, celebrates its premiere. A special materials pairing is used that requires no lubrication, ensuring long-term accuracy of the movement, especially the Swiss lever escapement.

2002

The magnetic field shielding developed in 1994 is further optimized. At the same time, the Tegiment technology is used for the first time in the Duochronograph 756. Hardened with this technology, the stainless steel case offers very effective protection against scratching.

2005

The birth year of watches made of submarine steel. This steel—sea water resistant and antimagnetic—is also used for the outer hulls of the German Class 212 submarine. German Lloyd of Hamburg certifies the diver's watch's pressure resistance.

At the Sports Press Ball in Frankfurt am Main in November 2005, Lothar Schmidt presents the SZ02 movement with a sixty-minute counter. It is the company's own development of a chronograph movement based on the Valjoux 7750. The movement is used in the football chronograph 303.

2006

The SZ04 watch movement, a conversion of the pocket watch caliber ETA 6498 to a regulator display, is unveiled. This design is incorporated in the Model Series 6100 Regulateur.

At the Dekra testing facility at the Lausitzring in Klettwitz, the Duochronograph 756 and the Multifunction Chronograph 900 undergo a crash test on the wrist of a crash test dummy. Both watches survive the torture undamaged.

Something new in the watch sector: Sinn diver's watches are treated like diving equipment and are tested and certified by Lloyd of Hamburg as per the European diving standards EN205 and EN 14143. The models tested are the U1, U2, UX, U200, U1000, and EZM3.

2009

Sinn celebrates ten years of Frankfurt Financial Center Watches with the Frankfurt Financial Center Jubilee II, which is limited to 100 watches, and the Frankfurt Financial Center in Platinum, which is limited to just ten watches.

2011

Sinn celebrates the fiftieth anniversary of its founding.

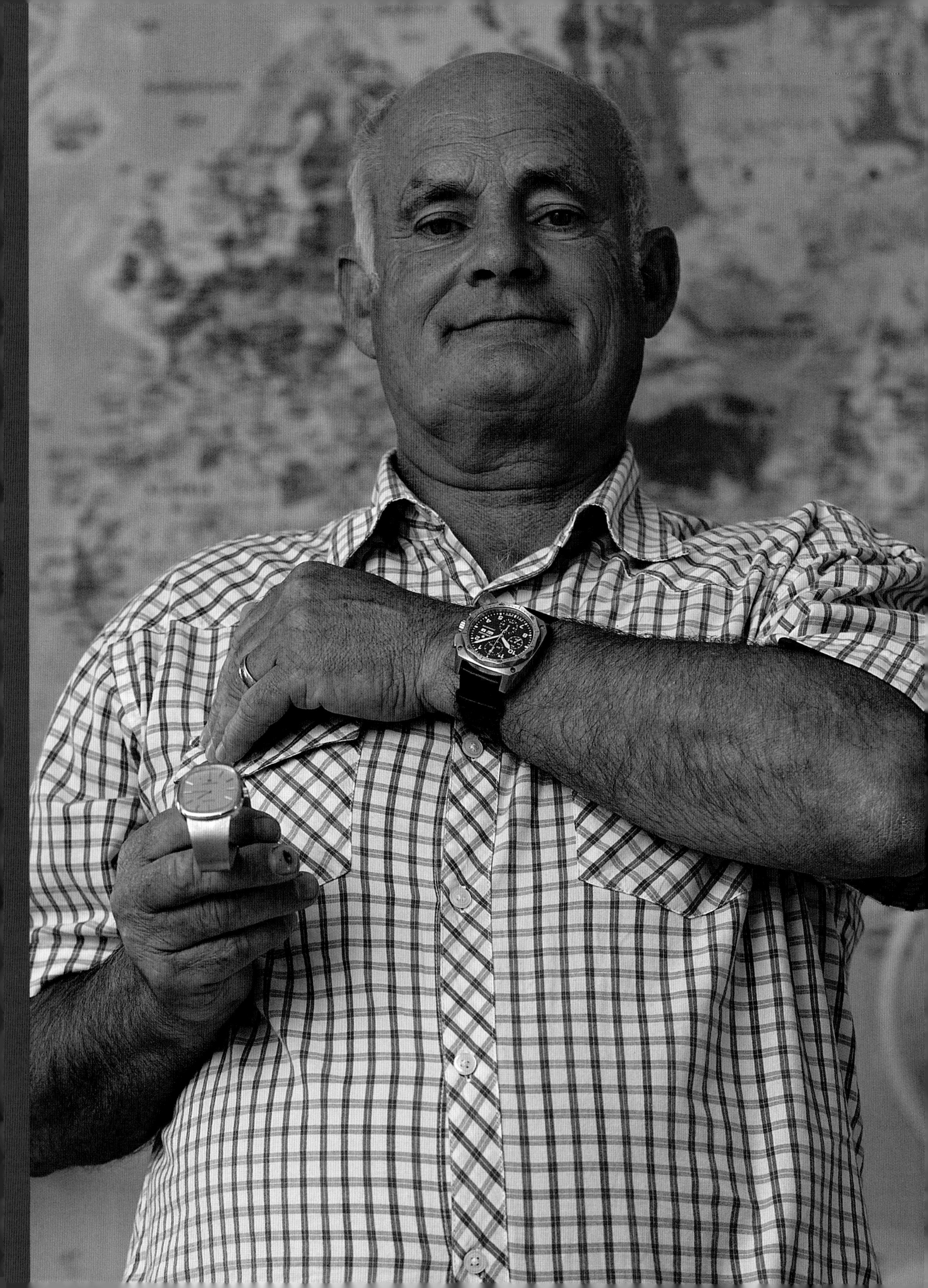

A Fateful Question about a Pilot's Watch

Jürgen Vietor was a pilot, heart and soul, first in Germany's naval air arm, and later with Lufthansa. In 1977, he experienced and survived the hijacking of the Lufthansa airliner Landshut by Palestinian terrorists, who actually wanted to shoot him—on account of his watch.

He is a man of contemporary history. He became so most involuntarily and in a highly dramatic fashion. When Jürgen Vietor took off from the Palma de Mallorca airport on 13 October 1977 as copilot on Lufthansa Flight 181, he had no inkling of the dramatic events that would make the population of Germany hold its breath for five days and place his life in danger on several occasions. Four Palestinian terrorists—two men and two women—hijacked the Boeing 737-200 in order to force the release of German terrorists of the Red Army Faction or RAF imprisoned in Stuttgart.

The leader of the terrorists was 23-year-old Zohair Yusif Akache, who chose to go by the name of "Captain Mahmud." Now 68, Vietor remembers him all to well: "Holding a pistol, he suddenly asked me in English, 'What is your religion?'" "Protestant," Vietor responded truthfully. Convinced he was lying, the terrorist screamed: "You're wearing a Jewish watch!" He had spotted a watch made by the German maker Junghans on Vietor's wrist. The Palestinian, who mistook its emblem, an eight-pointed star with a large J in the center, for a Star of David, forced Vietor to kneel in the aisle: "Now you die." Jürgen Schumann, captain of the Landshut, was finally able to calm the agitated terrorist and convince him that Vietor was a Protestant. That saved his life, but the watch was not so lucky, as Vietor related: "I was told to crush the watch under my foot, but with my crepe soles I was unable to do so because of the carpeted floor. Then Mahmud grabbed the crash axe from the cockpit and struck the watch repeatedly until the works fell out." Afterwards the terrorist joked with him as if they were best friends—an emotional rollercoaster.

Reminders of a decisive experience: artificial horizon, Boeing 737 model, and trim wheel.

This episode remains indelibly burned into his memory as do the dangerous landing next to the blocked runway in Athens, the brutal murder of his colleague Jürgen Schumann, and the subsequent odyssey, during which he had to fly the Boeing alone "with a dead captain on board and the mad Mahmud as co-pilot," as he sarcastically added. And finally there was the rescue of the hostages by the GSG 9 in Mogadishu. Two bullet-damaged parts from the Landshut—the artificial horizon and a trim wheel—serve as reminders, as does a watch. On learning of the story, the watchmaker Junghans gave him a gold wristwatch as a gift.

"I only wear it on special occasions," said Vietor when he showed me his treasure. Until recently, his everyday watch had been a simple quartz model, but then the retired pilot was given his choice of Junghans watches. In 2011, Junghans celebrated its 150th anniversary. To mark the occasion the company management invited Vietor, probably the bravest Junghans wearer, to the watch factory in Schramberg. As part of the formal ceremony Jürgen Vietor received the Junghans Chronoscope Pilot, which not only reminded him of a drastic event in his life but also took him back to the roots of his career as a pilot. The Chronograph Pilot is a modern interpretation of the old Junghans BW 111 chronographs, which was the service watch worn by pilots of the Bundeswehr. After the hijacking, Jürgen Vietor, who now lives in northern Hamburg, continued flying for Lufthansa, became a captain, "always on the 737," he stressed with a smile, and finally retired on account of his age in 1999. He still enjoys flying, but, as he adds with a mischievous smile, "only as a passenger."

As part of the Junghans anniversary celebrations, Jürgen Vietor (right) received a pilot's chronograph from Junghans managing director Matthias Stotz (right) and the owner of the company, Hans-Jochem Steim (center).

The Junghans Chronoscope Pilot is the modern version of the military chronograph that was the official service watch of the German Air Force in the 1950s. The military chronograph was equipped with the manually-wound Junghans Caliber J88. Its modern successor is powered by the ETA 7750 Valjoux automatic movement. Unfortunately Junghans removed this full-of-character pilot's watch from its catalog in 2012.

hanhart
1882
AUTOMATIC

Good Flight

Pilot's watches enjoy great popularity, and not just among pilots. As a rule, they are easy to read, have one or more additional useful everyday functions, and always give off a whiff of freedom. We have selected about a dozen new timepieces that splendidly combine functionality and modern aesthetics.

"Good readability is the alpha and omega of a pilot's watch," says Volker Thomalla. The man should know, after all, for as editor-in-chief of the leading aviation magazines *Aerokurier* and *Flugrevue*, he is in constant contact with aircraft makers as well as commercial and private pilots. And even with perfectly instrumented cockpits, all of these pilots value their personal timepieces, which make life in the pilot's seat a little easier. "If I am flying from Bonn to Stuttgart, of course I don't have to have a high-quality mechanical wristwatch," Thomalla admits frankly, but with a grin adds: "Being a pilot is a welcome excuse for me to put on a pretty watch." Pilots are simply no different than anyone else, and see a watch as a fine piece of jewelry. The motive of self reward is also present in aviation circles: "I bought myself my first pilot's watch when I got my instrument rating."

Even with all the irrationality affecting motives for buying and owning a pilot's watch, if it is a pilot's watch, it must also be functional. Readability has already been mentioned as important. "A second zone time is always an advantage, after all we pilots are always living in two time zones," says Thomalla. In addition to home time, UTC (Universal Time Coordinated) is always of importance to pilots, as it is used in scheduling flights and later recording them in the logbook. When a flight is over, and concentration and stress have dropped somewhat, it is easy to fall into the trap of wrongly calculating the difference in hours: "True. one notices this quickly, but it is simply useful to be able to read it directly from the watch." Thomalla prefers chronographs, especially when he is flying in an older aircraft: "I like the stop function if I am in a hold." That is what pilot's call the waiting patterns which, though less common, are still necessary from time to time. "This requires me to fly in a circle for precisely two or three minutes, and for this, I use the chronograph."

The aviation journalist would find a watch like the Sinn 900 Flieger S ideal. The 900, which represents the Frankfurt Special Watch maker's modern aviation line, has a starting price of about $3,400. With its scratch-resistant, black, hard-surface-coated case, it looks very cool, but its high-contrast dial and large subdials also provide the desired readability. An inner rotating bezel controlled by the crown at 10 o'clock allows for the accurate stopping of flight times, the Caliber ETA 7750 GMT offers stop-seconds, and a second zone time with 24-hour hand can also be adjusted in hour increments. The hour hand for local time, however, is fixed, which is always a minor drawback

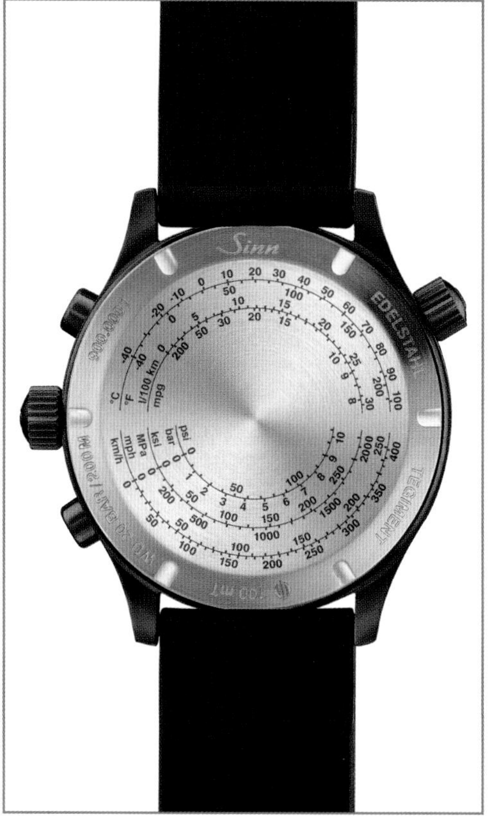

The Sinn Fliegerchronograph 900 S has a hardened stainless steel case with a black, wear-proof coating. Conversion tables for pilots are engraved on the case back.

on international flights. Otherwise, however, the Sinn is chock full of special aviation features: anti-fogging technology, magnetic field protection, pressure resistance to 20 bar, and five conversion tables for European and American units of measurement on the back of the watch case. This is made of nickel-free steel and the watch can thus be worn by most allergy sufferers.

Not designed specifically as a pilot's watch but entirely usable as such in keeping with the specified criteria, is the Chronoswiss Timemaster GMT (about $7,600). Because of its movement, the watch, which also has a black DLC coating, has the same functionality as the Sinn. At 44 millimeters, it is also exactly the same size, although the Chronoswiss looks a little bulkier on account of its large domed crown. However, the glass has a bi-directional rotating bezel with engraved 24-hour scale.

Completely designed to meet a pilot's needs is the Khaki Flight Timer made by Hamilton, part of the Swatch Group. This watch was developed in cooperation with pilots of Air Zermatt. An ETA quartz movement makes it possible to display more than time of day, second zone time, and date. Thanks to memory functions it can record taxi, takeoff, and landing times, making it a complete logbook for the wrist. The Hamilton has a starting price of about $1,450, and it was launched in 2012.

Oris also likes to have professionals among its clientele. In the aviation field the Hölsteiners work with the Swiss aerobatic pilot Don Vito Wyprächtiger,

While not designed as a pilot's watch, the Chronoswiss Timemaster GMT can be used as such. The red 24-hour hand indicates a second zone time.

who in 2010 became the first Swiss pilot to qualify for the Reno Air Races and took second place in this prestigious competition. In recognition of his achievement, Oris launched the BC3 Air Racing Limited Edition (about $2,160). The design of the watch is based a little on the Oris team aircraft. The tip of the 24-hour

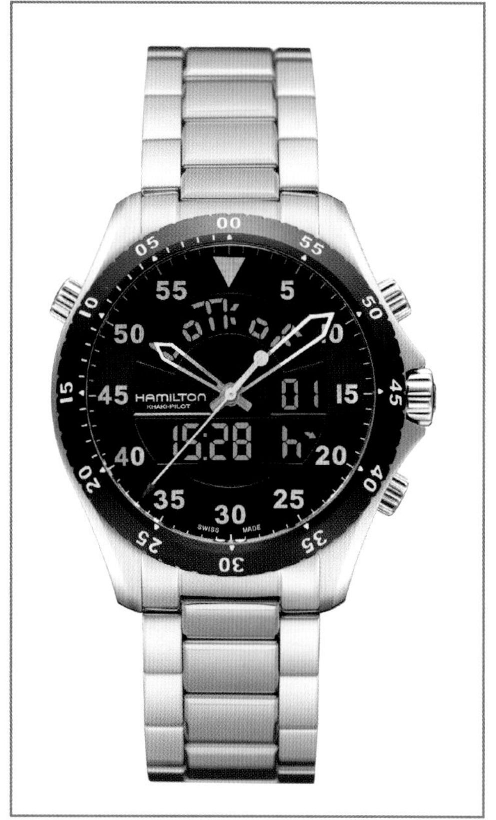

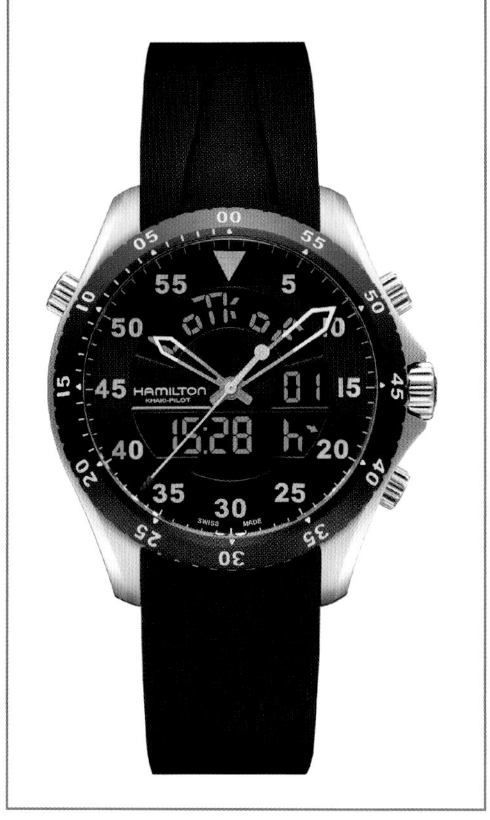

The Hamilton Khaki Flight Timer is literally a logbook worn on the wrist. According to Hamilton, this multifunctional quartz watch was developed with the pilots of Air Zermatt, whose helicopters fly air rescue missions.

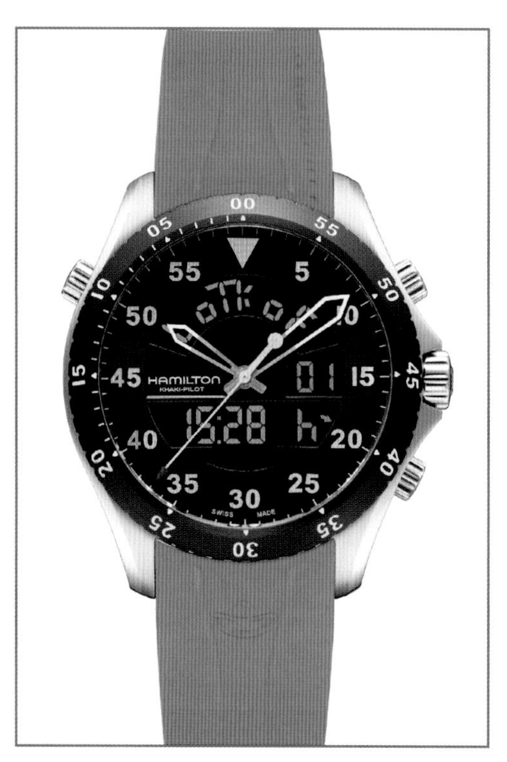

pointer is in the shape of an aircraft and is also painted red. Weight-saving measures were used, as in aviation. With a titanium case, it weighs only about half as much as a comparable steel watch. The large screw-down crown (BC = Big Crown) has massive flank protectors and is pressure resistant to 10 bar. The watch, which is limited to 1,000 examples, is delivered exclusively with a rubber and titanium strap and a strap-changing tool.

If Oris represents sports flying, then, at a price of about $11,000, the Senator Navigator World View by Glashütte Original can definitely be characterized as business class. It is a watch for Lufthansa gold card holders, who can also be called senators, at least at airports. A rotating inner bezel with imprinted city names informs the wearer of the respective zone times, while the readability of the panorama date, a typical feature of the

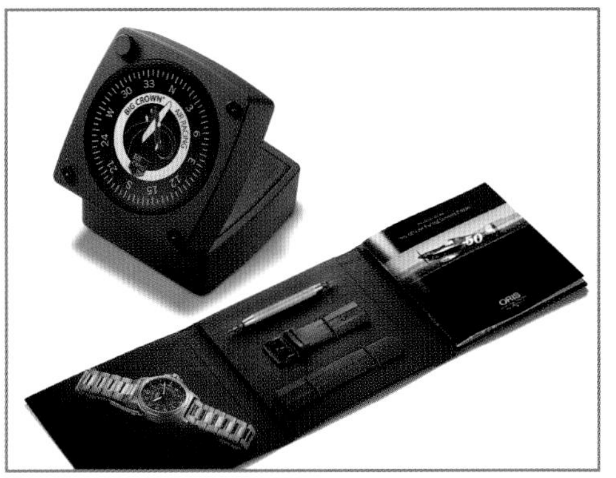

The Oris BC3 is a classically designed pilot's watch that has an indicator for a second zone time. The tip of the 24-hour hand is shaped like an aircraft. Viewed in profile, it quickly becomes apparent why this model series is called Big Crown.

Glashütte Original, can hardly be bettered. The classically designed watch is powered by an automatic winding manufacturer Caliber 39-47, which can also be admired through a glass case back.

In contrast, the Terranaut Trail by Glashütte's neighbor Nautische Instrumente Mühle is more aimed at the economy traveler. The steel model starts at about $1,200, while with a black PVD coating, the simple but extremely readable three-hand watch with date costs about $1,330, always with a calf leather strap. It is powered by a Sellita SW 200 automatic movement that is modified and improved by Mühle.

Because of its crown and pushbutton arrangement, the Chronograph Aerious by Erhard Junghans (beginning at about $4,400) has a much more unconventional appearance. The automatic Caliber J890, based on a reserved Seiko movement, was so aligned that the large start-stop button rests near the 3 o'clock position and the reset button between the bracelet lugs. Each may decide for himself whether this is practical, but in any case it is good looking. In addition to the ratchet-controlled chronograph function, it also has date and power reserve indicators. Such an attractive movement also cannot do without a glass case back. Versions in matte or highly-polished stainless steel are also available.

The buttons are also important characteristics on Hanhart chronographs. For one, the start-stop button is positioned slightly asymmetrically between 1 and 2 o'clock, Hanhart having modified the ETA 7750 base movement accordingly. For another, the reset button is often painted red, a feature with its origins in the early days of Hanhart aviation chronographs. This prominent red button first became a Hanhart feature in 1939. It was found to

Glashütte Original's pilot's watches are called the Senator Navigator. The World View model has a rotating scale ring with world time zones.

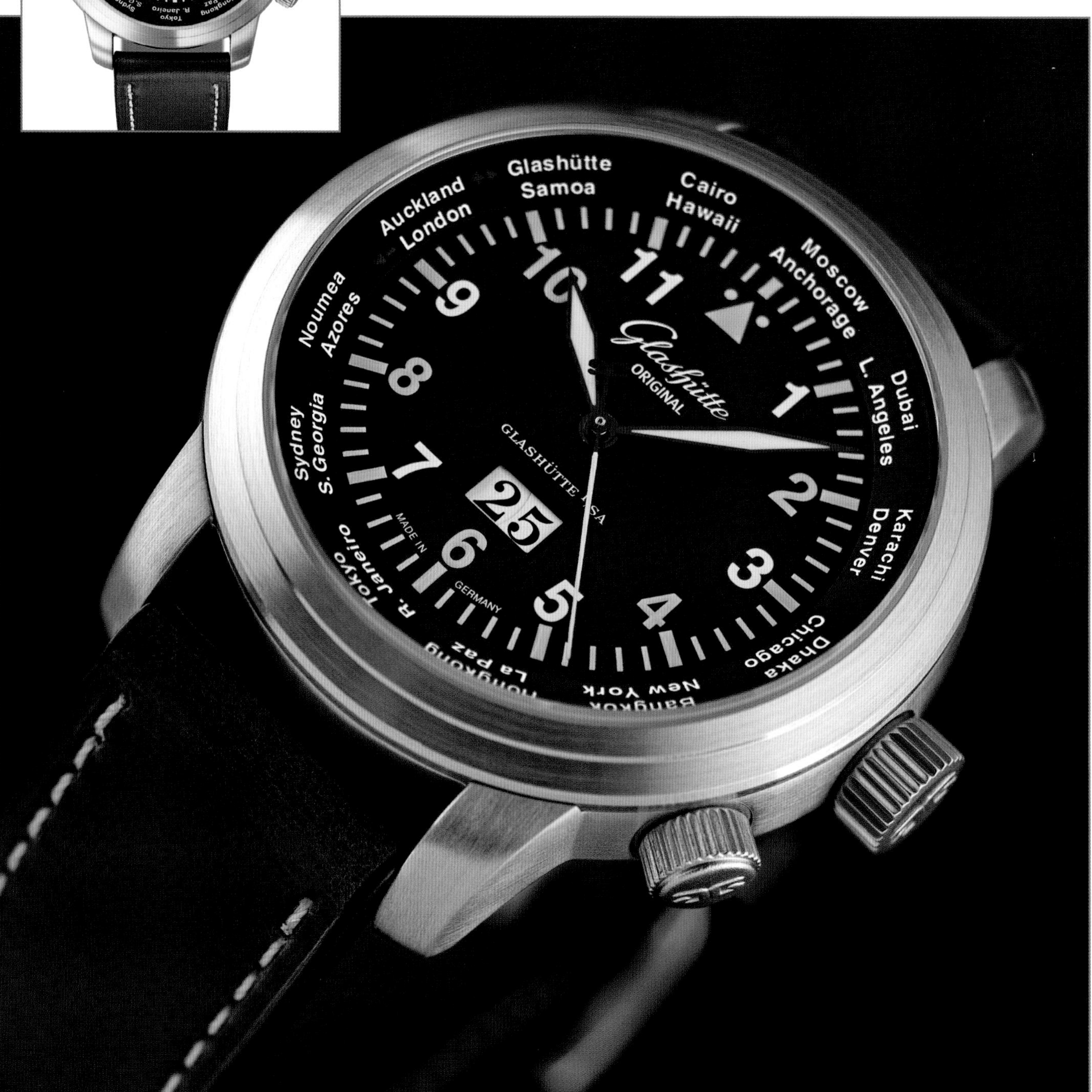

The Erhard Junghans Aerious is powered by the Caliber J890 chronograph movement, which is produced by Seiko exclusively for Junghans.

be an effective aid to pilots in navigation, preventing the inadvertent resetting of the stopped time, which could cause them to stray from the proper course. While today this is legend, it still looks good. In any case, with the Pioneer Twin Control, Hanhart has succeeded in transferring its history into a modern design.

Owner Peter Peter often has a hand or at least the last word in the design of Fortis watches; after all, he is a trained designer. The pilot's watches by Fortis are clearly designed, as one would expect. This is also true of the new B42 Flieger Big Date (starting price about $2,000). While its three hands are a classic design, its modern shape and typography are typical of a watch from the twenty-first century. In order to achieve balance on the dial, Fortis placed the weekday indicator at 9 o'clock and the large date window, which gives the watch its additional name, at 3 o'clock.

In its new pilot's watch collection called Startimer Pilot, Alpina, the sister brand of Frédérique Constant, also places emphasis on the contemporary interpretation of classical examples from its own house. In the 1930s, Alpina was known as the maker of robust timepieces and also made watches for military units in Europe. On this foundation is based the new collection in a handsome 44-millimeter case. It includes two three-hand watches (with date window and hand), whose prices start at about $2400. The top model is the Fliegerchronograph with Bi-Compax display (about $2,700). Alpina tried to smooth the way to this new market through cooperation with two renowned aviation companies: for the first two years, Cessna Aviation and Private Air were Alpina's advertising partners. An edition of 8,888 examples of the various models were delivered in gift boxes, one of which came with a model of the Cessna Citation Mustang in the colors of the airline Private Air.

Left: The Aerious Chronoscope is available in many versions. Here the classic pilot's watch design with highly polished stainless steel case may be seen.

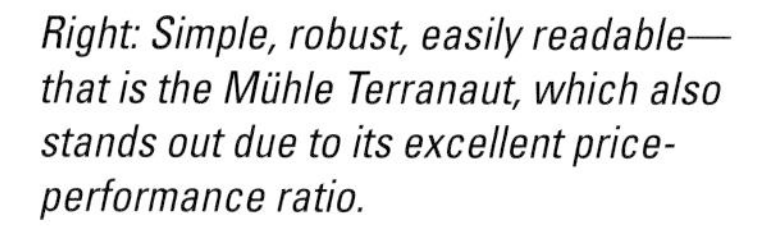

Right: Simple, robust, easily readable—that is the Mühle Terranaut, which also stands out due to its excellent price-performance ratio.

The Fortis B-42 Flieger Big Date is available with case in matte stainless steel or black coating. The arrows beside the date and day of week displays show the direction the crown must be turned to change the setting.

The new Alpina pilot's watch collection consists of three models, all in a luxurious 44-mm steel case.

While there are several modern examples here, Bell & Ross jumped quite far back into the past and took its inspiration from a pocket watch from the 1920s. The result was the new WW1 model, which stands for Wrist Watch 1. In the style of the old prewar pilot's watches, the Bell & Ross comes in a round steel case with round band lugs, although they are probably not still soldered on. So striking the exterior, so simple the technology. In the 45-millimeter case beats a simple ETA 2897 with power reserve indicator. The pilot's watch is accompanied by the PW1 pocket watch, powered by the hand-wound ETA 6497 movement. We can forgive Bell & Ross for bucking the trend towards modern pilot's watches, for after all, with the BR01 to 03 model series, the Franco-Swiss brand has a series of very modern, extroverted watches in its line.

A watch from the 1920s was the inspiration for the WW1 wristwatch and PW1 pocket watch by Bell & Ross.

Diver's Watches

IWC
PORSCHE DESIGN

Time Under Water

Whereas diving first became really popular about 30 years ago, diver's watches have been around for more than 70 years. One could write books about the development of diver's watches, chronographs, and diver's watches with alarms. We will at least dedicate a separate chapter to this theme, highlighting the history of diving watches with individual stories.

1927 may be considered the birth year of the watertight wristwatch. The young sportswoman Mercedes Gleitze swam the English Channel from Calais to Dover, a Rolex on her wrist. Its case design was patented and distinguished itself in part by a threaded case back that was screwed into the case and a screw-down crown. It was said to be as tight as an Auster (English oyster), which is why it was given that name. The crossing of the Channel was the first public demonstration of this watertight characteristic, which was promptly relayed in the newspapers on 24 November. "Rolex Oyster, the miracle watch which defies the elements", appeared on the front page of the Daily Mail. This was not entirely by chance, for Hans Wilsdorf, owner of Rolex, purchased space on the front page of the widely read paper for 40,000 francs.

As a result, the brand became famous overnight and has since been a symbol for the robust sports watch. The Rolex did not become a diving watch until much later. The Oyster case was systematically improved and formed the basis for the first Submariner, which became available in 1953 and was waterproof to a depth of 100 meters. Two Swiss scientists, Auguste and Jacques Piccard, who set out on their expedition equipped with Rolexes, played a significant part in its development and offered the watchmaker's engineers valuable tips. In 1960, Jacques Piccard helped Rolex gain further spectacular publicity by wearing an Oyster during his dive into the Marianas Trench. The watch was not worn on his wrist, however—a special design called Deep Sea Oyster, it was attached to the outer skin of the submarine Trieste. With its massive case and hemispherical glass, the watch withstood pressures in excess of one metric ton per square centimeter at a depth of 10,916 meters and remained absolutely tight. The mechanical movement performed flawlessly.

Testing Rolex watch movements under water predated Rolex diving watches, however. In 1936 the Italian company Guido Panerai and Figlio, from which Officine Panerai later emerged, designed the first prototype of a diving watch. The work was done on behalf of the Italian Navy, which had previously tested various Swiss watches but had not been satisfied with them. The Florentine engineers, who specialized in precise fine mechanics and built mechanical depth finders, compasses, and torpedo fuses for the navy, tackled the problem. Panerai installed a Rolex caliber in the case of a modified depth finder and that same year delivered it to the admiralty, which subsequently ordered ten examples. With the delivery of these watches in 1938, Panerai became the official provider of watches to the Italian Navy.

The dial was most unusual. It could be read in total darkness and consisted of two parts. The upper part was a black disc from which the large numbers and markers were punched out; the lower part was a dial completely covered with

This photo was seen round the world. In 1927, young sportswoman Mercedes Gleitze swam the English Channel—wearing a Rolex.

This Rolex is a one of a kind. Its name is Deep Sea Special Oyster and it dove to a depth of more than 10,000 meters with the submarine Trieste.

In its day, the Panerai Company of Florence made not just watches, but mechanical depth finders and compasses as well.

The original model of the Blancpain Fifty Fathoms. The hands and markers were covered with radioactive tritium. The yellow logo warns of radiation.

luminous material. A special mixture, which included zinc sulfate and radium bromide, gave the dial and hands their—in the truest sense of the word—special radiance.

The latter was also partly responsible for the name Radiomir and exposed its wearer to intense radiation—as if these soldiers' occupations weren't dangerous enough already. They were part of an Italian Navy frogman unit. Two divers rode on the back of a steerable torpedo, a sort of underwater motorcycle as it were, in order to sneak into Allied ports and destroy enemy ships with explosive charges. The most notable operation by the special-watch-wearing divers was in 1941, when they sank three large ships of the Royal Navy in Alexandria harbor. In 1943 the Florentine watchmakers designed a chronograph intended for naval officers. The watch bore the name Mare Nostrum, the old Roman term for the Mediterranean. The end of the Second World War meant that this project never got beyond the prototype stage, however.

In the early 1950s, Panerai delivered the Luminor model to the army and it later replaced the Radiomir. The shape of the case and the design of the dial remained the same, but the gamma-ray-emitting radium was replaced by less dangerous tritium. Its water resistance was also significantly improved, in large part because of the very special crown sealing, which to this day remains typical of a Panerai Luminor. It is a crescent-shaped dome over the crown, in which a small lever is installed. The lever has to be released in order to set the time or wind the watch. In the closed position it presses the crown sealing so hard that no water can enter and the crown cannot be moved.

Military requirements continued to drive development of the diving watch. In the 1950s, the French Ministry of Defense tasked Captain Robert Maloubier with the formation of a frogman unit, the Nageurs de Combat (Combat Swimmers). He not only had to find the right people, but the proper equipment as well. This included an extremely robust diving watch, which would be absolutely reliable and accurate at depths of almost 100 meters. Interestingly, the diving watch project took him to the watchmaker Blancpain in the Swiss Jura. It developed a diving watch which, in keeping with the requested diving depth, was called Fifty Fathoms.

A fathom, the traditional depth measurement used by Christian seafarers, is equivalent to 1.829 meters. Maloubier thus demanded a watch that was guaranteed waterproof to 91.45 meters and had extremely good readability and a dive time bezel, with which dive and decompression times could be measured accurately to five minutes. This made it an instrument with a safety role, with which the limited supply of air in the diver's tanks could be calculated. Maloubier also requested that the rotating bezel only move counterclockwise; the idea was to avoid situations where the bezel could be moved by accident, making it look as if the diver had been under longer and thus causing him to come to the surface sooner. Good readability was also part of Maloubier's specifications and was achieved by Blancpain through the use of large markers and hands that were coated with tritium. The watch underwent its practical test on 20 November 1953 and passed with flying colors. The French captain acknowledged the Swiss watchmaker's work with a simple "bon." Word of the Fifth Fathoms' reliability, quality, running accuracy, and ease of operation got round, and elite units such as the US Navy's SEALs and the recently established combat diver unit of the German Bundeswehr in Sangwarden also chose Blancpain.

The watch remained in official service until 1984. Then came the Ocean (p. 114), created by Porsche Design and developed by IWC on behalf of the Bundeswehr. The company in Schaffhausen won an official invitation to create a replacement for the Blackpain watch, and in the spring of 1980 ,it received thirty pages of specifications. The requirements included accuracy, shockproof rating, performance at various temperatures, pressure resistance, and anti-magnetic characteristics. The latter proved to be a real challenge, as Jürgen King, then IWC's technical director, recapped: "While we had experience with military watches, neither we nor any other company had ever made an anti-magnetic watch." The reason for this requirement: fuses of certain mines react not just to pressure or sound, but even to weak magnetic fields, such as those produced by the stepping motors of quartz watches. The new development could therefore not produce or change magnetic fields. The experienced watchmakers at Schaffhausen were forced to undertake some fundamental research.

And unconventional work. Because there are magnetic fields present in most buildings, King took the guinea pigs and testing equipment with him into his own garden and carried out the measurements there. Scientists at the University of Lausanne assisted IWC in development and procurement of materials, for the automatic movement was supposed to work like every other automatic movement, except that proven components had to be replaced by antimagnetic ones. One of these was the balance spring, which was made from a special alloy (Niob-Zirkon). Officially there is no further information, for the antimagnetic mine diving watch is automatic like a normal diving watch, and the combat diver watch with quartz movement remains in the navy's inventory to this day. It has a NATO supply number and is thus veiled in secrecy.

King can say one thing, however: "We were asked to build a very durable watch. But until we got there, there was a lot of scrap." The requirement for a robust case and glass resulted in a previously unattained level of waterproofness. The Ocean withstood 200 bar of overpressure, equivalent to a depth of 2,000 meters of water. As a result, both the civilian version of the Ocean and the military watches designed by Ferdinand Alexander Porsche got the name the Ocean 2000.

With the Blancpain already getting on in years and the IWC not yet ready, many combat divers turned to the first watch actually conceived for sport diving: the Doxa Sub 300. The unmistakable feature of a Doxa diving watch was and is its orange dial. Urs Eschle, former chief designer at Doxa, is regarded as the spiritual father of this dial. He dove in the lake at Neuchatel with four color charts (turquoise, yellow, red, orange), and because of its better readability at depth, he chose orange, the color of the 1970s. As a special feature, the designers provided the diving bezel with both minute graduations and a decompression scale.

Along with total submerged time, decompression is the most important time information for a diver. Scientist and extreme diver Hannes Keller realized this and, together with the developers at the watchmaker Vulcain, developed a new diving watch. The Cricket Nautical was the first diving watch with an alarm to have a decompression calculator, whose acoustic alarm also worked under water. The alarm signals the preselected surfacing time.

Keller had big plans, as he described in a letter to the Vulcain management on 3 July 1961: "I am happy to be able to report that I succeeded in setting a new diving world record in Brissago on the afternoon of 28 June 1961." Keller and his dive partner Kenneth MacLeish, science correspondent for the American magazine Life, had dived in Lago Maggiore, each

One of the first IWC Aquatimers from the 1970s. The dive time on the inner rotating bezel could be set using the second crown.

Doxa delivered its diver's series with orange-painted dials because it was believed that this made them more readable under water.

The Vulcain Cricket Nautical made diver's watch history as the first diver's watch with an alarm.

Sinn Special Watches had extreme diver Mario Weidner test its Arktis Diver's Chronograph (below) in the Polar Sea.

with a Cricket Nautical on his wrist. In his letter, Keller was full of praise for the timepiece, which endured "extraordinary stresses" such as high temperature and pressure differences: "Both watches passed the test with flying colors. I also tested the alarm mechanism of one of the two watches. The sound could be heard clearly, even though my diving suit was completely closed." Keller was also completely satisfied with the readability of the dial and the decompression table and reached the following conclusion: "In view of this extraordinary accomplishment, I have no doubt that the Vulcain Cricket Nautical diving watch will be a big success on the market and that the watch will serve many divers in the same way it served us during this difficult trial."

It was not to be the last time that an extreme diver took part in the development of a diving watch. On July 6, 1999, Mario Weidner set a record in the Polar Sea: from Frankfurt am Main, Weidner reached a depth of 64.5 meters in the icy water at the 81st degree of latitude. Over his thick suit he wore a diving watch that was also from Frankfurt: the Sinn-Chronographen 203. The professional diver later said, for the record: "The good underwater readability of the blue dial in conjunction with the slightly convex sapphire glass made a very positive impression." Sinn technicians who accompanied him determined that the special oil, used in practice for the first time, had lived up to its promise: despite the severe cold the automatic watch's movement did not deviate from the set point values. And the case did not let in one drop of Arctic water, which resulted in the Model 203 diving watch receiving the name affix "Arctic."

Meanwhile, Sinn also made diving watches whose cases were made of the same steel used in submarines of the German Navy. To underline the instrument character of their diving watches, Sinn

arranged for the models U 1000, U 1, U 2, and U X to be tested by Germanischer Lloyd—something like the Materials of Trade (MOT) requirements for shipping—and to be certified according to European diving equipment standards. Because a watch is not classic diving equipment, the EN250 and EN14143 norms had to be adapted. The testing was primarily devoted to accuracy and reliability at extreme temperatures. In one of the first tests, the timepiece spent three hours at -20°C (-4° F) , followed by another three hours at 50°C (122°F). Then, in a second series of tests, the watch spent three hours at -30° C (-22°F) followed by 70° C (158°F) and 95 percent humidity. German Lloyd subsequently tested pressure resistance, in the case of the U 1000 at 100 bar, equivalent to the pressure of 1,000 meters of water. The watch is thus capable of withstanding much more pressure than its wearer; however, he can be proud to own a watch with which one can casually swim the English Channel—or simply a local lake.

A History of Diver's Watches, Speeded Up:

1927: With the Oyster, Rolex unveils the first waterproof watch case. As public proof and to promote the innovation, Rolex gives an Oyster to the swimmer Mercedes Gleitze, who swims the English Channel wearing the watch.

1932: Omega makes its first waterproof and pressure-resistant watch, the Marine.

1936: Officine builds the first prototype of a diver's watch, with a Rolex movement in the case of a Panerai depth gauge.

1953: Blancpain launches the Fifty Fathoms, which is waterproof to almost 100 meters (50 fathoms = 91.45 meters). It is initially made for troops of the French Navy, and in 1959, it is also issued to combat divers of the German military. That same year, the Zodiac Sea Wolf is also released. It, too, is issued exclusively to a military user (US Navy SEALs). Finally, 1953 is the birth year of what is probably the best-known diver's watch in the world: the Rolex Submariner.

1954: The Rolex Submariner is certified to a depth of 200 meters.

1956: The Blancpain Fifty Fathoms becomes known to sports divers through the Jacques Cousteau film *The Silent World*. The Tudor Sub with Rolex movement appears in two versions, for depths of 100 and 200 meters.

1957: The Omega Seamaster is launched (waterproof to 200 meters).

1958: The Breitling Superocean, also waterproof to 200 meters, arrives on the market.

1960: To prove its capabilities, Rolex produces a single piece with an especially thick glass. Attached to the outer hull of Jacques Piccard's submarine Trieste, it survives a dive to 10,916 meters.

1961: The Swiss diving expert Hannes Keller dives to a record depth of 220 meters wearing the Vulcain Cricket Nautical watch with alarm.

1963: The Aquastar Bentos 500 is launched and accompanies the divers of Jacque-Yves Cousteau's team in the Précontinent-Experimente.

1964: The first diver's watches capable of reaching depths of 1,000 meters are unveiled. One is the reworked Aquastar/Lemania Benthos and the other the Caribbean 100.

1965: Doxa becomes the first brand to specifically go after the sport diver and, at the fair in Basel, unveils the Doxa Sub 300, the first diver's watch with an easy-to-read orange dial. Jaeger-LeCoultre presents the Polaris diver's watch with alarm, which has an internal diving bezel. Seiko presents the first Japanese diver's watch (waterproof to 150 meters).

1966: With the Favre-Leuba Bathy 50, Henry Favre brings to the market the first diver's watch with depth gauge after Boyle-Marriot. In the USA the watch is called the Bathy 160 (for diving depth of 160 feet).

1967: The bezel with decompression information developed in cooperation with the US Divers is patented on the Doxa Sub 300. In the USA, the Doxa is now also used by the military. Rolex makes a Submariner with helium valve for professional use (Comex). Seiko improves the water-resistance of its diver's watch and renames it Professional 300. IWC launches its first diver's watch, the Aquatimer.

1969: Doxa brings the 200T-Graph to the market, the first diver's chronograph with underwater stop mechanism.

1970: The new Omega Seamaster 600 accompanies three COMEX divers for eight days during saturation dives at 250 meters off Corsica. Diving pioneer Claude Wesley tests Doxa and Rolex prototypes with helium valves in the Red Sea. That same year Doxa introduces the Sub 600 Conquistador with helium valve to the market.

1971: Omega builds the Seamaster 1000. The Rolex Seadweller with helium valve (waterproof to 610 meters) appears.

1972: Omega introduces its first chronograph capable of use under water (waterproof to 120 meters).

1973: The Doxa Sub 250 appears, for the first time with the crown in the 4 o'clock position.

1975: Seiko doubles the diving depth of its diver's watch to 600 meters and for the first time uses titanium in the watch case. Also new is the rubber strap with three expansion folds.

1980: IWC receives a contract from the German military to develop and build a diver's watch. The Nautilus brand offers functional and reasonably priced diver's watches: the Professional for a depth of 500 meters for about 500 Marks and the Superpro with a depth of 1,000 meters for about 1,000 Marks.

1983: Sinn launches its first diver's watch, which is waterproof to 1,000 meters.

1984: The German military's development contract results in the IWC Ocean 2000, the first diver's watch in the world that is waterproof to 2,000 meters. It is made both in military (for the navy) and civilian versions. There is also an antimagnetic version for mine divers, which is waterproof to 300 meters.

1985: The Citizen Aqualand is the first analogue diver's watch with a digital electronic depth gauge. Seiko also introduces a diver's watch with electronic depth gauge, although it is much larger than the Citizen.

1986: TAG Heuer launches the Super Professional for a diving depth of up to 1,000 meters.

1987: Casio also unveils a diver's watch with electronic depth gauge, but in contrast to the Citizen, it only has a fully digital display.

1988: The Citizen Aqualand is further improved through the addition of a second display and a more compact sensor.

1989: The Rolex Submariner receives a rotating diving bezel that can only move counterclockwise. The Citizen Aqualand is now also available with an analogue depth gauge display. It thus becomes the most popular watch among sport divers.

1993: Omega designs a diver's chronograph for depths to 300 meters. In Lake Neuburger, free diver Roland Specker sets a world record of 80 meters while wearing the watch. In the same year, Omega launches the Seamaster Professional with helium valve, which is still in its catalog. The Seamaster appears in a James Bond film, thus becoming a "Bond watch."

1995: Breitling makes the Colt Superocean for a diving depth of up to 1,500 meters.

1996: With The Deepest, Kienzle breaks Rolex's world record. The record-setting watch is a simple plastic watch filled with oil for stabilization. It is certified to a depth of 12,000 meters. Sector unveils the Diving Team 1000 for depths to 1,000 meters.

1997: The Finnish manufacturer Suunto presents the Spyder, the first diving computer in wristwatch format, which becomes very popular among sports divers.

1998: IWC unveils the GST Aquatimer for depths to 2,000 meters.

1999: IWC launches the Deep One, the first diver's watch equipped with a mechanical depth gauge and drag pointer. It displays depths to 45 meters and is waterproof to 150 meters.

2000: Ulysse Nardin presents the first mechanical diver's watch with perpetual calendar.

2002: Breitling launches the Avenger Seawolf for depths to 3,000 meters and holds the current world record for mechanical watches.

2003: TAG Heuer launches the Model 2000 Aquagraph, the first diver's chronograph with stop function that remains functional at depths of up to 500 meters. Baume & Mercier (CapeLand S XXL 1000) and Oris (TT1 Master Diver) introduce watches with diving depths of up to 1,000 meters, while the new models from Bvlgari (Diagono Professional Scuba Diving 2000) and Mühle (SAR Rescue Timer) are tested to 2,000 meters. Doxa brings out the Sub 600 after the limited edition Sub 300 quickly sells out.

2004: IWC unveils the Aquatimer Split Minute chronograph. It is the first watch with a minute flyback mechanism that can also be used under water.

2006: Sinn has its U-series diver's watches certified to diving equipment standards by German Lloyd of Hamburg.

2007: Blancpain revives its legendary Fifty Fathoms model and subsequently introduces an entire model family under this name. Panerai presents the Luminor 1950 Submersible Depth Gauge, a diver's watch with mechanical depth gauge.

2008: Jaeger-LeCoultre presents the Master Compressor Diving Pro Geographic with mechanical depth gauge, waterproof to 300 meters.

2009: IWC introduces the Deep Two, the second diver's watch from Schaffhausen with mechanical depth gauge.

2011: Porsche Design introduces the Model P.6780 Diver with extendable movement cage.

ORIGINAL GAS ESCAPE VALVE
ROLEX
DEEPSEA
28
SEA-DWELLER
12800ft=3900m
SWISS MADE
10
20
30
40
50

Extreme Equipment

Rolex began the story of the really watertight diving watch with the Oyster case. The Submariner is one of the most popular diving watches on the market—and certainly the most copied. The most common impression left by the Deepsea, however, is that of a classic example of a hardcore diving watch.

This is an unforgettable watch. Anyone who has seen it knows what a diving watch is supposed to look like. Anyone who owns one will probably never leave it lying. The absence of just under half a pound on the wrist does not go unnoticed. And finally: Why should one take it off? Rolex has given the new diver's watch the very comfortable metal bracelet that earned much praise with the GMT Master II. The big watch made its debut in 2008, still under the name Dweller Deepsea, but for copyright reasons is now only called the Deepsea. Because Rolex pays attention to improvements in detail, the Deepsea was given an improved clasp. In Rolex nomenclature. this is called Glidelock. This stands for a patented system that makes it possible to adjust bracelet length with almost millimeter precision. Once adjusted, the watch fits even somewhat slimmer wrists—and with the help of the so-called "Fliplock," lengthening elements can also fit over a diving suit with a thickness of up to seven millimeters, or one-quarter of an inch.

With it, the underwater sportsman could—theoretically—reach the wreck of the Titanic on the bottom of the Atlantic. The Rolex, at least, would survive the dive undamaged, for it has been tested to 390 bar, equal to a diving depth of 3,900 meters. To withstand such pressure, even the Oyster case, acknowledged to be very robust, had to undergo some modifications. Among the visible changes are the slightly convex 5.5-millimeter thick sapphire glass and the titanium case back, also slightly convex. Both glass and case back, rest on a forged stainless steel ring, which is an integral part of the case frame. The Rolex designers borrowed this form of pressure load distribution from building statics, more precisely church-building, as a Rolex spokesperson acknowledged. Rolex calls the interaction of case back, glass, and support ring the "Ringlock System". Three gaskets see to it that no water enters through the tube of the screw-down crown. Each Deepsea must pass a water resistance test that Rolex developed in cooperation with the experts in underwater technology at COMEX (Compagnie Maritime d'Expertise).

Other improvements in the Sea Dweller concern readability. The golden hands and markers were enlarged compared to those of its predecessor and coated with blue luminous material, which can be seen better than the white under water. A point on the luminous surface points towards the triangle on the dive bezel that marks the zero point. The bezel itself has a ceramic coating, in which the numbers and markers are engraved and coated in platinum using PVD technology.

Finally, the Deepsea, like its predecessor, has a helium valve that underlines its professional appearance. Professional divers pause between dives at great depths in special capsules and breathe a high-pressure gas mixture that includes helium in particular. As a highly volatile gas, helium diffuses into the case during this time. When the diver surfaces, the pressure ratio changes and the helium is able to escape through the valve. Even ambitious amateur divers will scarcely ever use this function, but sometimes it is simply nice to know that one could do it if one wanted to.

Photo on facing page: Unobtrusive: the helium valve is discretely integrated into the watch case.

Double stitched: an additional steel ring reinforces the titanium case back.

Rolex also invested in the watch's inner value and improved the already robust Automatic Caliber 3135 with the high quality Parachrom helix, which is immune to magnetic field influences. The sum of these technical improvements is reflected in the price. The Deepsea can be had for about $14,000.

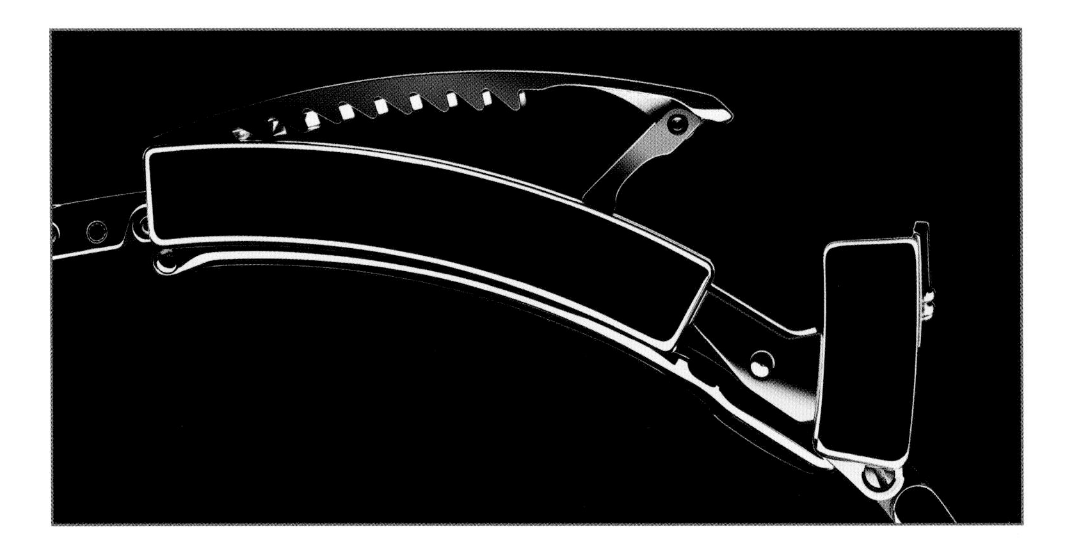

Refined: the "Fliplock" strap extension and the "Glidelock" fine adjustment are integrated in the latest clasp.

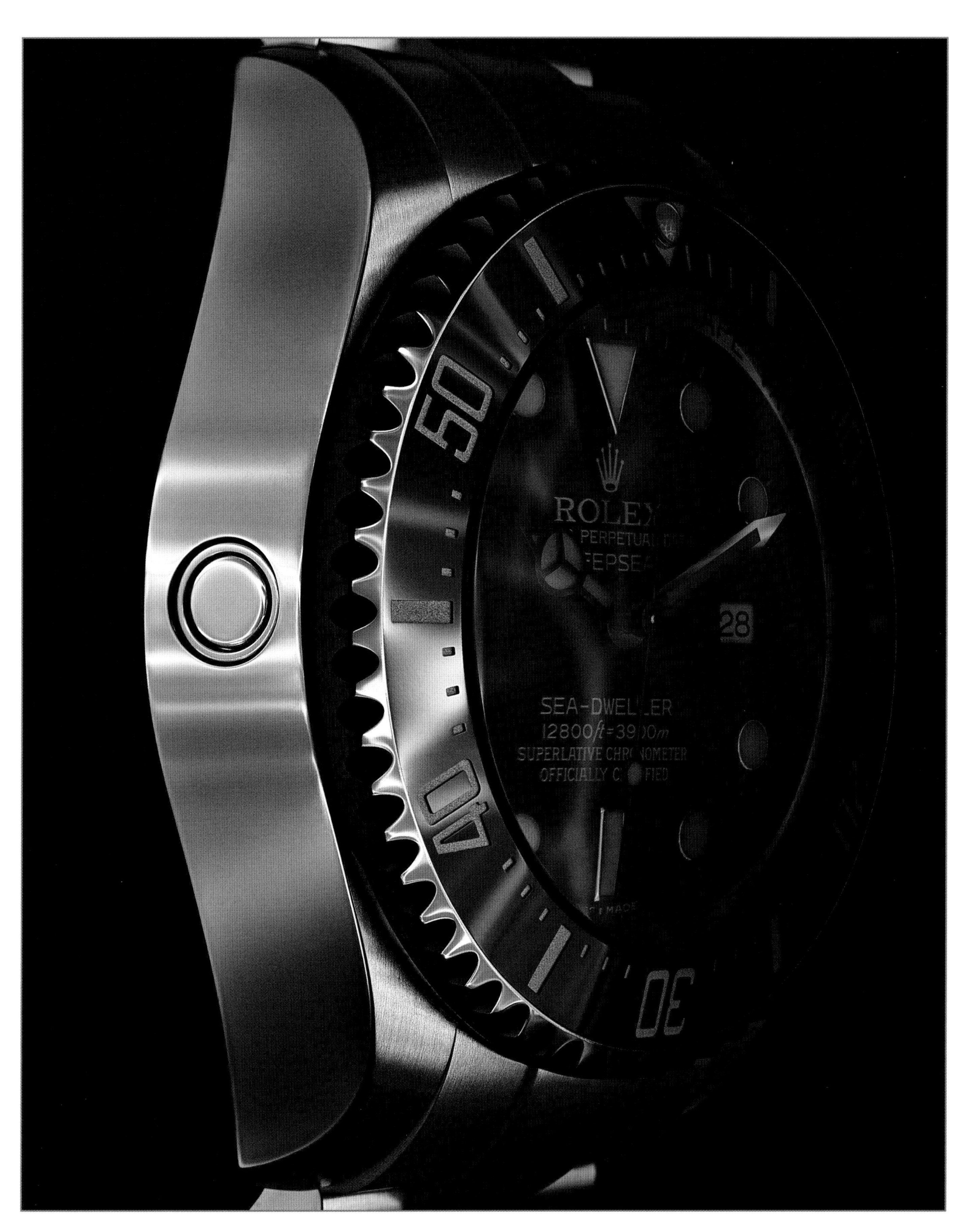
50
28
40
30

OMEGA
AUTOMATIC
Seamaster
600m/2000ft
PROFESSIONAL
OMEGA
CO-AXIAL
CHRONOMETER
1200 m / 4000 ft

Professional Divers

Waterproof watches are a tradition at Omega. In 1936, the Marine Model earned praise, and in 1957, the Swiss made the first Seamaster, which was deep-sea capable. And, just under 50 years ago, the first prototypes were built of a watch that was supposed to be "more watertight than a submarine."

The praise from the inventor of the bathysphere diving bell was highly personal. "It is water- and dustproof, its robustness and rust-resistance represent a significant advance in watchmaking," wrote Charles William Beebe on 23 June 1936 in a letter to Omega. With a Marine on his arm, the American dove several times to a depth of 14 meters and found that his Omega "successfully withstood the repeated stresses."

By about 20 years later, the stresses for divers and their watches were quite different. Exploration of the seas and their inhabitants, as well as their natural resources, took divers to ever-increasing depths. This trend also saw increased demands on the diver's watches, which at that time represented a sort of life insurance for the underwater worker. At that time, the state-of-the-art was represented by Rolex with its Submariner and Omega with the Seamaster 300, introduced in 1957, both of which were officially approved to depths of 200 meters. Here Omega's naming practice was not quite consistent. The control devices then in production and available to watchmakers could only be checked to a maximum of 20 atm of overpressure, which is why only 200 meters was guaranteed for the Seamaster. In laboratory tests, the watch stood up to significantly greater loads, which was documented in the name Seamaster 300.

Admittedly confusing for the customer, this did not bother the specialists: according to the Omega archive, in 1957 and 1958 deep-sea diving specialist Alain Julien made about 1,500 dives with his Seamaster, laying cables on the sea floor with colleagues or working in sunken shipwrecks, during which the watches were not spared. The Frenchman was quoted in the book *Omega: Journey through Time* released by the watchmaker: "The Seamaster 300 never displayed any kind of weakness, even after extremely harsh blows, in cold water and at great depths where this work was carried out. Thanks to their astonishing robustness, they remained reliable and accurate."

Omega again earned laurels in 1968, when two divers in a diving bell set a world record, reaching a simulated depth of 365 meters. Both wore Seamaster 300 watches. René Veyrunes and Ralph Brauer were employees of the Compagnie Maritime d'Expertise, or COMEX for short, which had reached a cooperative

The uninitiated may have a hard time distinguishing between the old and new Omega Ploprof. Here the current model is illustrated on the right.

The Marine established the tradition of Omega diver's watches.

agreement with Omega that same year. At this time, in Biel they developed a super diver's watch whose design was supposed to overshadow anything then on the market. The experiences of the COMEX people and those at the research center in Marseille were of great help to the developers in Biel.

The goal was a watch that could be guaranteed waterproof to a depth of 2,000 feet, or 600 meters. The engineers were not convinced that the classic three-part case design would be reliable enough. To reduce the number of possible leak sources, the Swiss developed a single-wall, so called monocoque, case, for which a patent application was submitted in 1967. The central case section and case back were one piece, literally milled from a solid block of steel. That made the watch relatively heavy for its time (approx. 150 grams). Case manufacturer Schmitz in Grenchen made several prototypes from titanium to reduce the watch's overall weight. At the time, processing the light, tough material was still complicated and costly, and consequently, plans for a titanium version were abandoned. Because of the watch case's monocoque construction, the movement was installed from the dial side. A chemically-hardened mineral glass was subsequently pressed onto a seal with 120 kilograms of pressure and then fixed in that position with a rotating ring.

Thus the winding shaft tube was left as the only potential leak source. Omega did not use the screw-down crown favored by its direct competitor Rolex. The sealing system used on the Panerai Luminor, in which a lever pressed the crown against a gasket, was also out of the question, as it had been patented in 1955 after many years in service. The engineers in Biel developed their own crown protection (patent application submitted on Oct. 23, 1968), which cleverly combined both systems: an integral, knurled nut in the central case section pressed the round extension of the square crown with a seal in the central case section. The crown disappeared almost completely into the massive flank protectors. The crown and flank protectors were located on the left side of the watch case at the 9 o'clock position to prevent them from pressing into the back of the hand and restricting movement of the wrist. This also left space for the prominent red button with which the bidirectional rotating dive ring can be unlocked.

The massive and clever design not only lived up to expectations with respect to waterproofing, it far exceeded them, as described in the book *Omega: Journey through Time*: "During a hydrostatic pressure test to determine the case's maximum strength, it ceased to run at a pressure of 137 atmospheres (equivalent to a depth of 1,370 meters). The reason was deformation of the watch case, which caused the glass to press onto the central second hand." A further investigation in the Ocean Systems Inc. diving research center revealed that, with their relative sizes taken into consideration, the Ploprof was "more watertight than a submarine." But being waterproof is no longer enough. Professional divers spend long periods under water using developments of Charles William Beebe's invention. They spend pauses between dives in a diving bell, where they breathe a mixture of helium and oxygen.

As a highly volatile gas, helium has the unpleasant characteristic of diffusing more easily than water in the watch case. Because pressure conditions change relatively quickly, even during controlled surfacing, the gas cannot escape quickly enough. This results in the creation of an overpressure situation inside the watch, which can blow off the glass under certain conditions. To avoid this phenomenon, many professional divers use a helium exhaust valve. Omega rejected this and decided to make the watch so tight that not even helium can enter. Tests with a helium leak detector confirmed this.

COMEX, which accompanied part of the four-year development program, immediately equipped its divers with the first production models. The Ploprof's baptism of fire came during a COMEX mission in the Gulf of Ajaccio. While searching for oil fields off the coast of Corsica, three divers remained at a depth of 250 meters for eight days. The deep-sea explorer Jacques Cousteau and his team used the new Seamaster in experiments at a depth of 500 meters, which ultimately confirmed the contemporary advertising slogan: "It may not look especially pretty above water, but deep below it is wonderful."

The Omega manager of the twenty-first century still believes in this statement. In a time when much in the watch market has become interchangeable and optional, a striking, unmistakable watch seems like a treasure to a product manager. That may well be the reason why Omega—as many makers have done with historical watches—has not produced a museum version of the Ploprof in small numbers. Instead it has included a modern interpretation of the classic, called the Seamaster Ploprof 1200, in its collection.

In a side-by-side comparison of the old and new Ploprof, the observer will be pleased to discover that Omega has created a modern watch without watering down the angular character of the original. The basic shape and size of the watch case have remained the same. Measured over the crown, it is 55 millimeters wide, while its length is 48 millimeters. The new watch's thickness has grown by 5 millimeters, however, to about 17 millimeters. This is due in part to the overall height of the modern Caliber 8500 with co-axial escapement, as well as to the fact that the new watch is pressure- tested to a depth of 1,200 meters.

The design of the case back is worthy of mention. It consists of two elements: a center section and a compression ring. The center section is positioned in such a way—with the aid of three lugs that fit exactly into corresponding recesses—that the case back with the seahorse medallion typical of Seamasters sits exactly level. More important, however, is its functional advantage: the O-ring seal can no longer be ruined by the turning of the case back; instead it is only compromised.

Although watertightness was nominally doubled, in terms of gas-tightness, the designers no longer rely solely on the watch case, which is why they gave the Ploprof a helium release valve. It is positioned opposite the unlocking button—now anodized orange—or roughly in the 4 o'clock position. The once patented crown protection system, which is no longer used on any other watch due to its complexity (read cost), has also been changed. The current Ploprof has a classic screw-down crown. It does, however, have a movable crown protector, which, in the closed position, joins positively with the watch case. Two integral and invisible tracks prevent the crown protector from turning when the crown is pulled.

The dive bezel has become clearly more modern and robust. While the system in the 1970 Ploprof was tough, it was made of the rather scratch-prone material Bakelite. The scale on the new model is coated with tough sapphire glass. The watch glass and dive bezel now form one surface, whereas on the old Ploprof the bezel projected a good two millimeters beyond the hardened and antiglare-coated mineral glass. The riffling of the new bezel is also clearly coarser, which simplifies operation while wearing gloves. Where there is light, there is also shadow: this coarse riffling is a very effective cuff destroyer. One must admit, however, that the Ploprof was not designed as a going-out watch.

Old and new: The crown of the collector's Ploprof is square. Its conical extension is pressed into a gasket in the case with the help of a knurled nut. The modern Ploprof has a screw-down crown, with a movable guard attached to its end.

Sapphire glass protects the diving bezel's scale against scratches on the current Ploprof.

Old and new II: The release button for the rotating bezel was originally made of plastic (above). Today it is made of steel and has an orange anodized ring.

The differences in the style of the hands are marginal. With a striking red-painted minute hand, an unobtrusive hour hand, and a central second hand, the new watch is almost identical to the old. But while the illuminated hour markers are printed right onto the dial of the historical model, on the new one, the luminous markers are applied. Finally, the date window has moved from the 3 o'clock position to one between 4 and 5 o'clock.

There remains the topic of comfort. Here a direct comparison would not be prudent, as, after all, the watches are equipped with different bracelets. The old Ploprof, which was made available to us by the Omega museum in Biel, was attached to the wrist by means of the original strap made of isofrane plastic. Despite—or perhaps because of—its age, this was quite stiff, which might not be a problem over a neoprene suit, but was uncomfortable on the naked wrist.

While the new version can be bought with a rubber strap, Omega sent the new Sharkproof, of which it is very proud, to the photo shoot. As we discovered, their pride is justified. For the thick Milanese strap is a true skin flatterer, and with the help of a folding clasp, fits the wrist perfectly. The heavy watch doesn't shift even on slender wrists, but with its total weight of about 270 grams (a little over half a pound), it lives up to the name given it by some watch fans: "diving weight with time indicator."

The Bond Watch

Professionals trust Omega, both in real life and in films. This is why the Seamaster Professional has an unofficial name affix known even to nondivers: the Bond Watch. The models current at the time have, after all, undergone difficult trials on the wrist of Agent 007.

The Seamaster made its first appearance in a Bond film in *GoldenEye* with Pierce Brosnan as 007 in 1995. It had two refinements in its case: a laser that shot from the helium valve, and a remote detonator for bombs, which after activation triggered a green light on the dial at 12 o'clock. Bond used the laser to cut open the floor of the armored special train in order to escape from Alec Trevelyan (the renegade Agent 006, who also wore a Seamaster in the film—albeit an older model). The detonator was used later when it was supposed to arm a bomb to destroy the secret satellite activated by GoldenEye for the Cuban command central.

In the Bond film *Morning Never Dies* of 1997, the Seamaster Professional again had a detonator, with which a shell on the stealth boat of media mogul Elliot Carver, the film's villain, was set off in order to make it visible to radar.

The watch also played a role in the 1999 Bond production *The World Is Not Enough*, as a gadget designed by master weapons designer "Q." A powerful light source emitted from the dial, giving Bond an overview of the situation from his inflatable jacket-airbag, in which he found protection from an avalanche. In another critical situation, the agent used an arrestor hook hidden in the case to free himself and Dr. Christmas Jones from a bunker: the miniature arrestor hook shot from the crown and caused the Seamaster's bezel to spin rapidly. As soon as the hook had a firm hold, the bezel slowly turned in the other direction, winding the wire onto the watch case and pulling Bond and his companion to safety.

Then the inventiveness of Her Majesty's weapons maker drops drastically—at least with respect to watches: in the last Bond appearance by Pierce Brosnan, the Omega had the same added functions as in his first: the laser and bomb detonator. Daniel Craig, as a less humorous, tougher Bond, even had to get by without special features in *Casino Royale, Quantum Solace* and *Skyfall*. For him the Seamaster Professional has been what it represents in real life: an elegant, robust sports watch.

James Bond relies on nothing more than his Omega Seamaster Professional, which cuts a good figure even with a black suit.

LUMINOR
PANERAI
Automatic
24
15
30
45
REG.

Under and Over

The Panerai Luminor Submersible is a monumental diver's watch that not only stays tight under water but also cuts a good figure above it.

Panerai polarizes. The died-in-the-wool mechanical watch lover will find the self-assertive appearance of these watches too ostentatious, while those to whom inner value is not as important will treasure precisely this look. But essentially both factions are entitled to their opinion, at least as far as all Panerai models with in-house movements are concerned. Its full name is Luminor Submersible 1950 3 Days Automatic—which itself betrays much about its characteristics. It is part of the Luminor series, its shape is based on that of a model from 1950, and its automatic winding movement has a power reserve of 72 hours, or three days. A Submersible is a small underwater vehicle. It is, therefore, a diver's watch.

One special feature is the crown seal, typical for a Luminor since the 1950s. It consists of a crescent-shaped protective frame over the crown, in which a small lever is installed. The lever must be released to set or wind the watch. In the closed position it presses the crown seal so powerfully that no water can enter and the crown becomes immovable.

The manufacturer claims that the watch is waterproof to 30 bar, which is equivalent to a water depth of 300 meters. We never made it that deep during practical tests in the Caribbean Sea, but for swimming and snorkeling, the Panerai was a reliable partner, seemingly unaffected by salt water and sand. Of course after each use we rinsed it carefully with fresh water, which we also recommend after swimming in chlorinated water. The Submersible has all the ingredients of a diver's watch: first there is the dive bezel, which can only be rotated counterclockwise and in minute steps and which has an illuminated mark at the zero setting and minute markers to 15. It is attached to the wrist by a real rubber diver's strap and a massive titanium pin buckle. We have already mentioned its water-resistant qualities.

Quite unusual for a diver's watch is the viewing window in the case back. Panerai includes this on all models equipped with a manufacturer movement. In this case it is the Caliber P.9000, which was unveiled at the Geneva Watch Salon SIHH 2010. It displays the time of day—with the typical Luminor small seconds subdial at the 9 o'clock position—and the date. A minor specialty is the separately adjustable hour hand. If the crown is pulled to the first stop, for example ,while changing the time zone, the hour hand can be moved forwards and backwards in hour steps without stopping the watch. The date can also be corrected in this way. With an average loss of two seconds

The crescent-shaped lever presses the crown seal into the case.

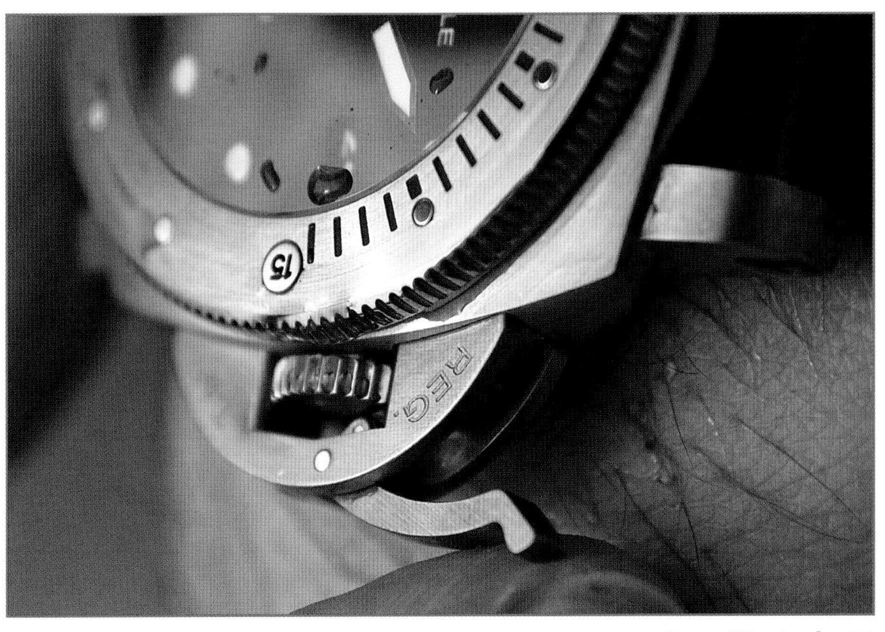

per day, the movement operates within the norms for chronometers, which is an outstanding value. Personally, however, we would ask the watchmakers, to find an easier method of adjustment. Two spring barrels connected in series provide a running autonomy, from fully wound to discharge of the mainspring, of 72 hours. In practice, this means that the watch can be left on the night table for two days without worry; the self-winding mechanism will provide the necessary power supply when the watch is worn.

Panerai is one of the brands which, in keeping with its tradition, pushes lavish watch diameters. The smallest is the Radiomir Chronograph at 42 millimeters. The Submersible measures a proud 47 millimeters—and is 16 millimeters thick. It is therefore a real whopper, which fits under a shirtsleeve only with difficulty. But thanks to the use of titanium, the watch is not overly heavy and, in combination with a skillfully designed case, provides real wearing comfort—even on more slender wrists. Panerai asks close to $10,000 for the Titanium-Submersible, which, given the market situation for a diver's watch with manufacturer movement, is within reason. In return, the buyer receives not only a reliable, well-finished timepiece but also a great deal of attention as well—see above.

Neither salt water nor sand left any visible marks on the test watch.

Because Panerai is proud of its manufacturer movement, it includes a viewing window in the case back of its titanium diver's watch.

A robust pin clasp and a rubber strap with expansion member are convincing proof that the Submersible wants to be a true diver's watch.

VULCAIN
-300m-6-1000f-
VULCAIN
-300m-6-1000f-

Diving Cricket

Long before diving computers provided important information, underwater sportsmen used mechanical watches to go down and come back up again safely. A top instrument in the 1960s was the Vulcain Nautical with integrated decompression computer and underwater alarm.

Total time under water and decompression times represent the most important time information for divers. The first is limited by the amount of air in the dive tanks and must be adhered to accurately to avoid suffocating under water. The latter indicates the intervals a diver must adhere to in order to avoid so-called diving sickness, which in extreme cases, can lead to death. The modern underwater sportsman deals with both of these tasks, including depth measurement, with a highly precise dive computer.

In the 1960s, such devices did not yet exist, and mechanical solutions were sought. And so scientist and extreme diver Hannes Keller, together with developers of the watchmaker Vulcain, devised a new diver's watch. The Cricket Nautical was the first diver's watch in the world with an integral decompression calculator, whose automatic alarm functioned under water. The alarm was supposed to signal the preset surfacing time.

Bernard Fleury is the CEO of Vulcain, where the legacy of Maurice Ditisheim is preserved without fuss but with great commitment.

The instrument was basically an invention of Robert Ditisheim, founder of the Swiss watch brand Vulcain. He took the cricket as his model and then used the name for his new product. The tiny animal uses chirps to alert other members of its species about its territory from several meters. Ditisheim therefore reached the conclusion that it must be technically possible to also produce a very loud noise in a wristwatch case. He displayed proof of this when the Vulcain Cricket was unveiled in 1947. Ditisheim quickly discarded the concept of sound springs, as they failed to produce the desired results. Instead, he used the previously mentioned cricket as a model and experimented with the insect's method—amplification of sound through resonating bodies—with a double base. The inner sound base, which also closed the case, was responsible for producing the rattling noise. Made to vibrate by a hammer, it produced sound waves. These were amplified by the resonance space between the inner and outer bases, causing the Cricket's alarm to sound loudly and distinctly, whether it is lying on the night table or attached to the wrist of a diver.

As previously mentioned, the first wristwatch with an alarm in the world was also the first diver's watch with alarm—

The vintage Cricket Nautical offers a look at its inner workings. Beside it are the inner part of the back, made of brass, and the steel case back. The resonance space between the two creates the penetrating noise.

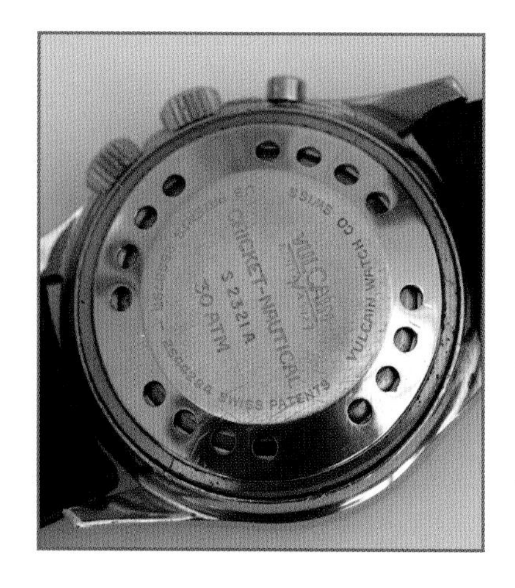

Ready for use: the Cricket Nautical with case back in place.

with the name affix Nautical. The Caliber 120 was transplanted into a watch case capable of withstanding diving depths of up to 300 meters. Devised by Hannes Keller, the decompression table in the center of the dial was nothing less than revolutionary for its time. The crown at 4 o'clock is used to rotate the disc in either direction. A dial window that turns with it displays information about decompression times. Even though there are probably few divers who still rely on information from a mechanical watch, in the owner's manual there is still a warning to the buyer of a new Nautical: "Attention, your watch is an exact replica of the original model made in 1961. Information in the decompression tables has changed since then. Please consult a diving specialist, in order to check the current norms."

The replica is not quite an exact one, but the changes are actually of a cosmetic nature. Whereas the dial of the original was black, that of the replica, which appeared two years ago, is dark blue. The strap of dark blue rubber with functional folding clasp is matching and contemporary. The original, which was made available to us by colleague and Vulcain specialist Michael Horlbeck, is attached to the wrist by a so-called "tropical strap"—a plastic webbing strap with pin clasp.

Even though the movement, which consists of 157 individual parts, was given a new caliber name—it is now called V-10 instead of the previous Caliber 120—it is completely identical in design and function. It has two barrel springs for the clock and alarm mechanisms, although both can be wound using the same crown. If the crown

A decompression table is printed on the dial of the classic Nautical.

Old and new: In profile, the Vulcain logo may be seen on the crown of the current Nautical (below).

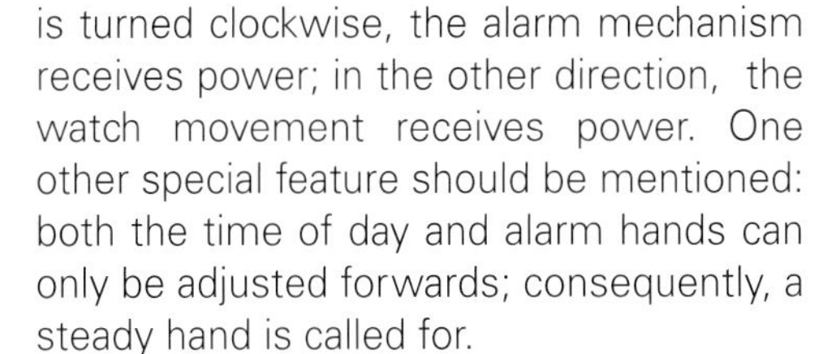

is turned clockwise, the alarm mechanism receives power; in the other direction, the watch movement receives power. One other special feature should be mentioned: both the time of day and alarm hands can only be adjusted forwards; consequently, a steady hand is called for.

The question is always asked: original or replica? And the answer is always the same: that is a matter of taste. For about $4,150, the purchaser of the replica gets a well-finished manufactured product which, given the price trends for complicated mechanical watches, can be characterized as a good buy. Those desiring a whiff of history, however, will have to pay at least $1,400 more. And take care when buying. For as Vulcan expert Horlbeck knows, there are a few phonies among these much-sought-after collectors items.

Vulcain Nautical

Reference: 100107.064RF/BE

Movement: Manually wound, Vulcain Caliber Cricket V-10, diameter 28 mm, thickness 5.6 mm, 17 jewels, 21,600 A/h; second mainspring barrel for alarm function.

Functions: Hours, minutes, central seconds, alarm.

Case: Stainless steel, diameter 42 mm, thickness 17.4 mm; Hesalite glass, double resonance case back; waterproof to 30 bar.

Bracelet: Rubber with double folding clasp.

Versions: Pink-gold case, various strap and dial models.

Old and new II: While the classic Nautical came with the "tropical" fabric strap and pin clasp, the new one has a rubber strap with folding clasp.

ZALIUM-PLATINUM
AUTOMATIC CHRONOGRAPH

Harry Goes into the Water

Harry Winston Rare Timepieces is known as a maker of complicated dress watches lavishly loaded with diamonds. The Geneva maker also makes a diver's watch—the Project Z2 Diver. In doing so, the brand fully lives up to its name. Almost everything about it is rare: Production is limited to twenty watches, and the material used in the case, Zalium, is unique.

The location is well chosen. The legendary underwater film *The Big Blue* had been shot off the coast of Sicily. And in Taormina, on the island's east coast, a notable film festival with illustrious guests takes place annually. The sea and high society are ideal ingredients to reveal to the public a diver's watch from a noble brand. Harry Winston is undoubtedly a noble brand. It is not unusual for a piece by the New York jewelry makers to go for more than a million dollars. And this, according to New York representative Susy Korb, is far from being the upper limit. Although it now sounds rather decadent, a watch like the elegant Excenter Timezone for about $50,000 can almost be considered a bargain. When the Z2 was unveiled in 2005, Maximilian Büsser, current owner of the manufacturer MB&F, was still managing director of the Geneva company Harry Winston Rare Timepieces. At the end of his career with Harry Winston, Büsser, who significantly advanced the exclusive jewelers' watch division, recalled his start with the company: "When I started, the average price for a necklace was $150,000, and we made watches for 20,000 Swiss Francs [about $23,000]." Watches by Harry Winston have become a little more expensive; the highly complicated models of the Opus series, made in small numbers, are even more expensive. But what Büsser said at the time has not changed much even today: "The watches provided an entry into the brand environment of Harry Winston."

The Project Z2 Diver was the first serious diver's watch from the luxury brand Harry Winston. The case is made of the company's own Zalium alloy.

The Z2 adopted the characteristic crown guard of the Z1, the first sports watch from Harry Winston.

That is also true of the Project Z2 Diver. When it was introduced, the asking price for the diver's watch was about $25,000. To loyal Harry Winston customers, therefore, it had almost the character of an accessory. The edition, limited to 200 watches worldwide, quickly sold out. A special feature of this watch is its case made of Zalium. Harry Winston copyrighted the name, which stands for an alloy whose principal component is zirconium. This metal is very resistant to acids and alkalis and is also highly corrosion-resistant, which predestined it for use in aviation and rocket-building. Because of its high resistance, it transfers no harmful substances to the human body and is therefore frequently used in the manufacture of surgical instruments.

Because of its extremely high density (6.49 g/cm^3), zirconium is even tougher than titanium (4.5 g/cm^3), although it is heavier and more difficult to work with. For example, the case center section cannot be punched out—the norm in watchmaking. In the truest sense of the word, it must be milled from the solid using special hard tools. This increases costs, as does the fact that zirconium, while very commonly found on the moon, is rather rare on earth. It is therefore ideal for a "Rare Timepiece," especially since Zalium, developed by chemist Ronald Winston, is used exclusively for the company's own brand. The company stubbornly refuses to reveal the exact composition of the alloy. Max Büsser once remarked with a smile: "Coca-Cola has never revealed its recipe to this day."

A Harry Winston product will never become a massed-produced article like Coca-Cola, which was Ronald Winston's intention when he founded the watch division in 1988. At that time, it was estimated that 5,000 watches were sold per year in 32 countries. These included the successor to the Z2, the Ocean Diver with Zalium case and a bezel in either white or red gold—both typically cost about $62,000.

With this diver's watch the designers achieved something that advertising writers extol, even when nothing else pleases them: sporting elegance. This watch is the exception, however, hitting the nail right on the head. The sheer size of the case (44 mm diameter) exudes sportiness, as does the fact that the Z2 is a chronograph. The choice of materials guarantees elegance. The grey, semi-matte Zalium case is combined with a rotating bezel, crown, and pushbutton made of platinum. The buyer can choose between rubber or crocodile leather for the strap, depending on his preference for sportiness or elegance.

The shape of the case is consciously based on that of the Z1. The first sports watch by Harry Winston was unveiled at BaselWorld 2004 and built in a limited production run of 100 watches. From it, the Z2 adopted classic features like the crown protector and movable strap lugs. The latter can be tilted downwards up to 30 degrees from the horizontal and see to it that this less than delicate sports watch fits comfortably even on slender wrists. The two guards on either side of the crown not only protect its flanks, but also bridge, in the true sense of the word the diving bezel, which can only be turned counterclockwise. The chronograph button, which thanks to its large area is easily operated, fits organically into the curve of the case.

The slate grey dial is also somewhat different than the usual. The permanent seconds subdial, which is reduced to a pure function control, is particularly striking.

The Ocean Diver model is the successor to the Project Z2 Diver. Both examples shown here have a Zalium case. The watch on the left has a red-gold bezel, that on the right has one made of white gold. Price: in the $45,000 range.

The seconds subdial is shaped like a Japanese throwing star and is blue tempered.

Inside it turns a *shuriken*, a miniaturized and tempered replica of a Japanese throwing star. Even though Asian customers might be flattered, the weapon symbol on a watch worn by a European seems at least questionable. The throwing star—like the hands, numbers, and markers—is generously provided with luminous material, ensuring good readability even under poor light conditions. This is also true of the chronograph minute and second hands, which stand out against pale grey dials.

By the way, zirconium alloys are also used to shroud fuel elements in nuclear reactors. In the figurative sense, this is also true of the Z2 Diver, but fortunately the contents of the Zalium case only radiate beauty. The watch is driven by the very refined automatic chronograph Caliber 31C6 by Girard Perregaux. The owner of the watch will see nothing of this movement however. It is hidden behind a massive Zalium case back with engravings. This helps make the watch waterproof to 200 meters and—thanks to all relevant features—also allows it to be characterized as a true diver's watch. The only question is whether anyone would really use this watch for diving; but perhaps for a short dip in the sea of Taormina.

Harry Winston Project Z2 Diver

Movement: Self-winding, Girard-Perregaux Caliber GP 31C6; diameter 26 mm; thickness 6.3 mm; 27 jewels, 28,800 A/h.

Functions: Hours, minutes, central seconds as function indicator, date, chronograph.

Case: Zalium, diameter 44 mm, thickness 16 mm; bezel rotatable in one direction with 60 graduations; screw-on case back; screw-down crown; waterproof to 200 m.

Bracelet: Rubber with Zalium pin clasp, crocodile leather strap available.

They Dive a Little

Diver's watches are not a trendy idea, but rather an ongoing success story. These sporting instruments for the wrist do more than cut a good figure under water. Here are several current models.

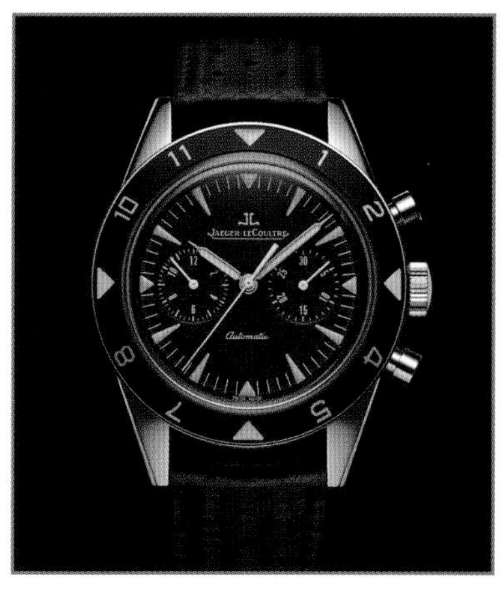

It is a fact that no one actually needs a diver's watch to go diving these days. They have largely been replaced by the diver's computer, which can show the exact decompression time as required, even if the diver changes depth several times in the course of the dive. Nevertheless, diver's watches continue to be popular. One professional diver explained: "There are more sport divers now than ever before. But few are interested in going to work, to a restaurant, or to church with their bulky diver's computer on their arm." On the other hand, there is a desire to outwardly demonstrate one's membership in the diving community, and for this the diver's watch is perfect. Finally the diver's watch is a practical everyday companion—waterproof, robust, and good-looking. The spectrum of current diver's watches ranges from modern entry-level models to established classics, from the simple stainless steel watch to pink gold chronographs. There is something for everyone.

When divers talk about watches the name Rolex always comes up, for the Submarine model has written an almost incomparable success story. In the process the Submarine has become self-propelled. Special models like the so-called "Frogmariner" with the green dial have become collector's items, and the stainless steel version of the diver's watch (about $7400) is so sought after that it is not always in stock at every outlet. Lovers of classic diver's watches also strike it rich in Rolex's sister brand Tudor. The Heritage Black Bay model recalls a successful Tudor model from 1954. This watch was on the market until the 1980s and had a pressure rating of 20 bar (200 meters), making it a real deep diver. The case diameter was increased to 41 millimeters to suit modern tastes without destroying the vintage look, which is characterized by loving details such as gilded hands and the burgundy-colored bezel. The Tudor comes with a steel and fabric strap and costs about $3,000.

At IWC, a military contract was the catalyst for the development of diver's watches. The company from Schaffhausen received an official contract from the Bundeswehr, and in the spring of 1980, began developing the Ocean model. The military laid down its requirements in a 30-page specification. These included accuracy, shock resistance, temperature behavior, water resistance, and antimagnetic behavior. This know-how flows into the current diver's watch collection with the name Aquatimer. The top model is the Deep Two (about $16,200), a steel watch with integrated mechanical depth gauge to 50 meters of water. The latter also has a drag pointer so that both the actual and maximum diving depths can be displayed. The IWC diver family also includes the Aquatimer 2000 (about $4,300), a classically-formed diver's watch that is pressure resistant to

From top to bottom: Rolex Submariner, Tudor Heritage Black Bay, various versions of the IWC Aquatimer, Jaeger-LeCoultre Memovox Deep Sea.

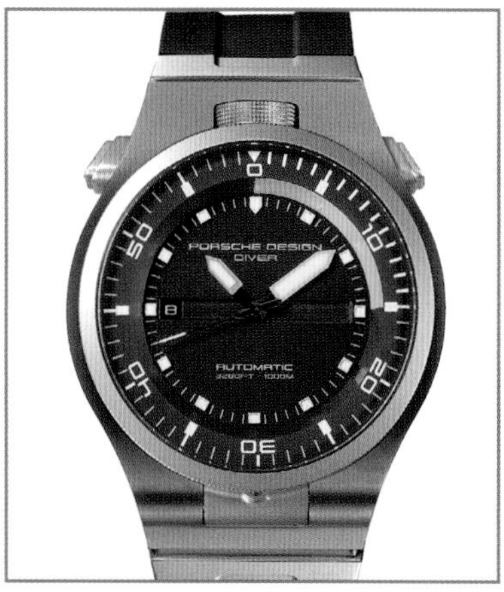

From top to bottom: Ulysse Nardin Blue Sea, Tutima DI 330 Black, Oris Tubbataha Limited Edition, Porsche Design P.6780 Diver.

200 bar (2,000 meters). Both are driven by an automatic movement which IWC designates with the caliber number 31110 and is based on the ETA 2892-A2.

Jaeger-LeCoultre, like the IWC part of the Richemont group of companies, also had a hand in writing the diver's watch story. When diving became popular as a sport in the 1950s, the Swiss made the Memovox Deep Sea, a watch with an alarm to remind the diver it was time to surface. The Deep Sea Vintage Chronograph is based on this alarm wristwatch. While a chronograph is not particularly useful under water, as it is almost impossible to read, this does nothing to diminish the watch's other qualities. Behind the elegant, classic shell is the modern manufacturer Caliber JLC 751G with ball-bearing-mounted winding rotor and large balance wheel, which shines on account of its accuracy. This chronograph, which is waterproof to 10 bar (100 meters), is always suitable for swimming.

Ulysse Nardin has been making marine chronographs for precise time measurement on the water for more than 160 years. Sport divers can dive with the Blue Sea wrist chronometer without worry, which is symbolized by the wave marking on the dial. The hour markers and hour and minute hands have blue luminous surfaces for good readability. The date is displayed in a window at 6 o'clock. The watch has a bezel that rotates in a single direction and a screw-down crown, and its case is waterproof to 200 meters. Production is limited to 999 examples, with each watch numbered on the side of the case.

No less elegant, but slightly cheaper at about $1,400, is the DI 330 Black by Tutima. As the name says, the classic diver's watch comes in all black, and the light titanium case is PVD-coated. In combination with a rubber strap and a titanium clasp, the result is a light, comfortable watch that displays not only the time but the day of the week and date as well.

Because the minute is the most important unit of time to most divers, in its diver's watches Oris employs a regulator dial with large central minute hand and decentralized hour hand at 3 o'clock. The Tubbataha Limited Edition also has this feature. Its outstanding features include the blue dial, the blue ceramic inlay in the bezel, and a special case back decoration. It shows a shark, representative of all the threatened species in the Indonesian Tubbataha Riff National Park, which benefits from sales of the watch, which sells for about $2,600 with rubber strap.

The P.6780 Diver by Porsche Design is impressive both under water and at the bar. This is not just because of its technical look but also because of a special functionality: the time and rotating diving time bezel can only be adjusted if the steel movement container is raised from its titanium bridge structure after the two release buttons on the side have been pressed. Not only is this a safety feature, but it is impressesive to the observer as well. Depending on his taste, the buyer can have the watch in natural finish or with PVD coating. In both cases, the jeweler's asking price is about $8,000.

IWC
AQUATIMER
DEEP TWO
ACT. DEPTH
MAX. DEPTH
300 PSI
P MAX 20 BAR
EN250